Tamed

Alice Roberts is an anthropologist, writer and broadcaster, and is currently Professor of Public Engagement in Science at the University of Birmingham. She has presented several landmark BBC series including *The Incredible Human Journey*, *Origins of Us*, *Coast* and *The Celts*. Her latest book on evolutionary biology, *The Incredible Unlikeliness of Being*, was shortlisted for the Wellcome Book Prize in 2015.

Praise for *Tamed*

'Lyrical storytelling untangles the current thinking on how we've entwined our lives with those of plants and animals. From dogs to apples to potatoes to chickens, Roberts provides fascinating insights into domestication, offering anecdotes from past and present that link genetic and archaeological findings.'

BBC *Wildlife Magazine*, Book of the Month

'Superb: fascinating, intimate biographies of the species that have shared our white-knuckle ride to the present and have helped to make us what we are. Read if you want to know what and why you are.'

Charles Foster, author of *Being a Beast*

'Roberts remains composed, engaging and undogmatic throughout ... *Tamed* is an excellent point of entry for anyone who wants to understand the new deep human history.'

Peter Forbes, *Guardian*

ALSO BY ALICE ROBERTS

Tamed

Ten Species That Changed Our World

ALICE ROBERTS

✸ WINDMILL BOOKS

1 3 5 7 9 10 8 6 4 2

Windmill Books
20 Vauxhall Bridge Road
London SW1V 2SA

Windmill Books is part of the Penguin Random House group of companies
whose addresses can be found at global.penguinrandomhouse.com.

Penguin
Random House
UK

First published by Hutchinson in 2017
First published in paperback by Windmill Books in 2018

www.penguin.co.uk

A CIP catalogue record for this book is available from the British Library.

ISBN 9781786090010

Typeset in 12/15 pt Sabon MT Std
by Integra Software Services Pvt. Ltd, Pondicherry

Printed and bound in Great Britain by Clays Ltd, St Ives Plc

Penguin Random House is committed to a
sustainable future for our business, our readers
and our planet. This book is made from Forest
Stewardship Council® certified paper.

To Phoebe and Wilf, who love the wild places

Contents

Introduction

'HEAR *and attend and listen; for this befell and behappened and became and was, O my Best Beloved, when the Tame animals were wild. The Dog was wild, and the Horse was wild, and the Cow was wild ... and they walked in the Wet Wild Woods by their wild lones ...'*

Rudyard Kipling, 'The Cat That Walked By Himself'

For hundreds of thousands of years our ancestors existed in a world where they depended on wild plants and animals. They were hunter-gatherers – consummate survival experts, but taking the world as they found it.

Then the Neolithic Revolution happened – at different times, and in different ways, in different places – but across the globe, those hunter-gatherers were changing how they interacted with other species in a crucial fashion. They tamed those wild species – and became herders and farmers. The domestication of plants and animals would pave the way for the modern world – allowing the human population to boom, and the first civilisations to grow up.

By uncovering this deep history of familiar species, we'll discover just how important those plants and animals were – and are – to the survival and success of our own species. These others have teamed up with us, and are now found right around the world, and have changed our lives immeasurably. We'll dig back in time to trace their – sometimes surprising – origins. But we'll also find out how becoming part of our world changed those plants and animals, as we tamed them.

The origins of domesticated species

When the Victorian scientist Charles Darwin set about writing *On the Origin of Species* – the foundation stone of evolutionary biology today – he knew he was about to drop a bombshell – and not just into the world of biology. He understood that he had to prepare some serious groundwork before he leapt into explaining his extraordinary insight into how species changed over time, through the unthinking action of natural selection, working its magic, generation by generation. He needed to bring his readers along with him. They'd be climbing a mountain together; it would be fraught with difficulty, but the view from the top would be stupendous.

And so Darwin refrained from jumping straight into explaining his revelation. Instead, he devoted an entire chapter – a whole twenty-seven pages in my edition – to describing examples of species evolving under the influence of humans. Within a population of plants or animals, there is variation – and it's by interacting with that variation that farmers and breeders are able to modify breeds and species, generation by generation. Over hundreds and thousands of years of humans promoting the survival and reproduction of some variants, and limiting the success of others, our ancestors had wrought change in domesticated species and strains, moulding them until they more neatly fulfilled human needs, desires and tastes. Darwin called the effect of human choice on those domesticated species 'artificial selection'. It was an idea that he knew his readers would be familiar and comfortable with. He could describe how selection by farmers and breeders – picking out particular individuals to breed from, discarding others – would, over generations, produce small changes, and that these changes would accumulate over time so that sometimes diverse strains or subtypes would emerge – from a single, ancestral stock.

In fact, this gentle introduction to the power of selection to wreak biological change wasn't just a literary device. Darwin had set out to study domestication himself, because he believed that it could cast light on the mechanism of evolution more generally – on how wild plants and animals could become gradually modified. He wrote, '… it seemed to me probable that a careful study of domesticated animals and cultivated plants would offer the best chance of making out this obscure problem. Nor,' he added, almost with a glint in his eye, 'have I been disappointed.'

After discussing the effects of artificial selection, Darwin could then go on to introduce his key concept of natural selection as the mechanism behind the evolution of life on the planet, the unthinking process that would, over time, propagate modifications and grind out – not just new strains – but entirely new species.

Reading his book today, the word 'artificial' trips us up. Firstly there's the other meaning of 'artificial' – where it's synonymous with 'fake'. That wasn't the sense in which Darwin was applying the word; he meant artificial as in 'by artifice'. But even then, there's a knowingness implied by this word which overplays the role of conscious intent in the process of domestication of species. Modern plant and animal breeding may be carried out with careful, deliberate aims in mind, but the earlier history of our liaisons with the species that have become our major allies reveals a shocking lack of any planning.

So we could try to come up with a new word for 'artificial', but there's another problem. Given that we now accept the fundamental role of natural selection in evolution, given that Mr Darwin doesn't need to persuade the majority of us of this biological reality – do we actually need a separate description for the way that humans have affected the evolution of domesticated species?

Describing artificial and natural selection separately helped Darwin to build his argument, and to introduce a challenging, new idea, but the distinction is actually false. It doesn't really matter that it's us humans – rather than the physical environment or other species – that are mediating the assortment of individuals into those more or less likely to successfully reproduce. You wouldn't make this distinction for any other species. Take the selective pressure exerted by honeybees on flowers – which leads to changes in those flowers over time, making them more attractive to their pollinators. The colours, shapes and scents of flowers are not designed to delight our senses – they have evolved to entice in their winged allies. Have the honeybees been effecting artificial selection? Isn't this just bee-mediated natural selection? Perhaps, when it comes to our own influence on domesticated species, instead of 'artificial selection' it's better (although, admittedly, slightly clunkier) to think of it as 'human-mediated natural selection'.

Natural selection works its wonders by weeding out particular variants, while others survive and reproduce – passing genes on to the next generation. Artificial or 'human-mediated natural selection' often works in the same way, as farmers and breeders reject certain plants or animals which aren't as docile, productive, strong, tall or sweet as others. Darwin described this negative selection in the *Origin*:

> When a race of plants is once pretty well established, the seed-raisers do not pick out the best plants, but merely go over the seed-beds, and pull up the 'rogues' as they call the plants that deviate from the proper standard. With animals this kind of selection is, in fact, also followed; for hardly any one is so careless as to allow his worst animals to breed.

By pulling up the rogues, sifting out the animals that we don't want to breed on, or even just looking after certain animals better than others, humans have become powerful agents of natural selection. We've roped in a great variety of plants and animals to become our allies in the game of life.

Yet, as we'll see, sometimes this taming seems to come about almost by accident. And sometimes it appears as though the plants and animals are actually domesticating themselves. Perhaps we're not as all-powerful as we once believed ourselves to be. Even when we're setting out deliberately to break in a species, to make it more useful to us, we're really only unlocking a natural, latent potential to *be* tame.

The deep histories of plants and animals that are very familiar to us today take us to strange and exotic locations. It's a good time to be tracing these stories. Arguments have raged on over how each domesticated species sprang into being – from a single origin, a single discrete centre of domestication, or from a wider geographic area, as different wild species or subspecies were tamed and then interbred to form hybrids. In the nineteenth century, Darwin thought that separate wild species could explain the immense variety we see in our domesticates. In contrast, the great, early-twentieth-century plant-hunter and biologist Nikolai Vavilov thought that he could pinpoint discrete centres of origin. Archaeology, history and botany provide plenty of clues, but also leave us with plenty of unresolved questions. With the arrival of genetics – a new historical source – on the scene, we now have a hope of testing competing hypotheses and solving these seemingly intractable puzzles, to uncover the real story of the plants and animals that have become our allies.

The genetic code carried by living creatures contains within it not just the information to make the modern, living organism, but also traces of its ancestry. Looking at the DNA of living species, we can delve into their deep past – thousands, even millions of years back – and glean some clues. We get further insights if we can add genetic clues from DNA extracted from ancient fossils. The first contributions from genetics focused on small fragments of genetic code, but in just the last few years, genetics has broadened its scope to look at entire genomes, and has produced a panoply of surprising revelations about the origins and histories of some of the species closest to us.

Some of those genetic revelations challenge the way that we divide up the biological world. It's useful – and meaningful – to identify species. That concept embraces a group of organisms that are diagnosably similar to each other – and diagnosably different to others. But the fact that populations undergo evolutionary change, over time, can make drawing boundaries around species quite difficult. We do like to put things in boxes, but biology seems to delight in breaking itself out of such constraints, as we shall learn again and again in this book. How far do lineages have to diverge before they are truly separate species? That's still a question which taxes taxonomists. When it comes to domesticated animals and plants, some are considered to be subspecies of their wild counterparts, and are given the same species name as their untamed progenitors, and surviving wild cousins – if there are any. Some biologists have advocated using entirely separate species names for domesticates, even if they're very similar to wild relatives, for ease of reference. The debate over naming shows just how blurred the borderlines are.

In each case, the evolutionary trajectory of the domesticated species – from cattle and chickens to potatoes and rice – was profoundly influenced by becoming intertwined with that of an African ape that had already spread across the world and gone global. These stories are extraordinary and manifold, but I have focused on just ten species. One of those species is us, *Homo sapiens*. The astonishing transformation we've undergone – from wild apes to civilised humanity – suggests that we have somehow tamed ourselves. And only after that happened could we set about taming others. I leave the story of humans to the last chapter. There are many surprises and very new revelations there – hot off the scientific press – but you'll have to wait for those. Spend time first with nine other species. Each one has had a huge impact on us and our history – and is still important to us today. These domestications are scattered through time and space, so that we'll understand how human societies have been interacting with plants and animals in various ways, around the world and through history. Their spread around the globe accompanies our own movements – sometimes even fuelling and propelling those human migrations. We find dogs running with hunters; wheat, cattle and rice travelling with the earliest farmers; horses carrying their riders out of the steppe into history; apples stowed in saddlebags; chickens spreading with empires; potatoes and maize crossing the Atlantic on trade winds.

The Neolithic, which first started around 11,000 years ago in East Asia and the Middle East, formed the foundations for the modern world. It was the most important development in the entire history of humanity. We became entangled with other species – in symbiotic relationships that meshed our evolutionary paths together. Farming created the capacity to grow the global human population to immense proportions. Our population is still growing, but we're pushing at the boundaries of the capacity of this planet to support us. We need to – quickly – develop sustainable ways of feeding at least a billion or two more of us than already live on earth.

Some solutions may be low tech – organic farming has proven much more promising than its detractors suggested, even just fifteen years ago. But high tech may form part of the solution, too. We need to decide how we feel about embracing – or rejecting – the newest generation of genetic modification – tools which can deliver precise genetic adjustments to suit our needs, bypassing the selective breeding our ancestors relied on – or even creating new possibilities that are limited only by our imaginations.

There are other challenges – with a human population that's still growing, and four-tenths of land already farmed, we need good evidence to show us the best solution for preserving as many wild species as possible. We're clever – that's always been a characteristic of humans. But we need to be cleverer than ever if we're to find a way of balancing the voracious appetite of a growing human population, and the hordes of tame species we need to survive, with biodiversity and real wilderness. It can sometimes feel as though we humans are a plague on the planet, and it would be a complete catastrophe if the real legacy of the Neolithic Revolution was mass extinction and ecological devastation. We have to hope that there might be a greener future for us – and our allies. Scientific research may not only illuminate the history of our interactions with other species, it provides us with powerful tools to inform the future directions that we can choose to take. Knowing more about the histories of our domesticated species will help us plan for the future.

But let's start with the past, and see where that takes us. We're going back far into prehistory to begin the journey, to a world unrecognisable today. A world with no cities, no settlements, no farms. A world still in the chill grip of the Ice Age. Where we meet the first of our allies.

1

DOGS

Canis familiaris

> *When the Man waked up he said, 'What is Wild Dog doing here?' And the Woman said, 'His name is not Wild Dog any more, but the First Friend, because he will be our friend for always and always and always. Take him with you when you go hunting.'*
>
> Rudyard Kipling, 'The Cat That Walked by Himself'

Wolves in the woods

The sun had set and the temperature had dropped even further. These were the cold, hard months, when the day was so short that there was barely enough time to hunt, to mend the tents, to chop wood for the fire. The temperature outside never rose above freezing. Towards the end of winter, things always got difficult. The dried berries from last summer would eventually run out. Then there was meat for breakfast, meat in the middle of the day, meat for supper. Mostly reindeer meat, of course. But occasionally, just for a change, a bit of horse or hare.

There were five tents in the camp: tall and conical, like robust tipis. Each was built on a skeleton of seven or eight larch poles, covered with hides, all stitched together and tied in place against the wind. Under the snow, a ring of stones held the skirt of the tent down. The fallen snow, at least half a metre deep around the sides of the tipi, also helped to keep the hides secure. Between the tipis, the snow was trampled. The remains of a hearth lay in the centre. It was barely used now – during these frozen weeks, it was much better to light fires inside the tents. And so, in each one, a fire blazed in a central hearth. The contrast in temperature was extreme. As the families retreated into their tipis for the night, fur coats and trousers and boots were discarded in a great pile by the door.

Outside the ring of tipis was a place for chopping wood. One or two men would split felled larch trees all day, enough to keep the fires in the tents burning. In another place lay the scant remains of what had been a reindeer. It had been butchered into pieces, and there was little left apart from a few ribs and blood-stained snow. The hunters had killed it that morning, and brought it back to the camp. When they arrived, they had immediately opened up its belly to take slices of still-warm liver to eat, and to drink its blood. The rest was divided up amongst the five families, and carried off into the tents. Apart from the head – once the tongue and cheeks had been removed, the antlered skull had been carried back into the edge of the forest. A young man had taken it, tying it to his belt and climbing several metres up a larch tree before wedging the skull between a branch and the trunk: a sky burial; an offering to the spirits of the forest and the spirit of the reindeer itself.

After another meal of mainly meat, the families started to settle down for the night. Children were tucked up under piles of reindeer skins. The last adult to go to sleep in each tent stacked logs onto the fire. It would burn on for another hour or two. Then the temperature inside the tent would drop, almost to the ambient chill outside. But the reindeer fur would keep them warm, just as it had kept its original owners warm through the icy winters in this cold, northern land.

As the skeins of blue smoke escaping from the tops of the tents grew thinner, and the murmur of conversation died down, that meagre carcass on the edge of the camp drew scavengers out of the forest. Emerging from the taiga – like shadows, skulking and silent – the wolves approached the camp. They made short work of the remains of the deer, and then they prowled around the tents and the central hearth, searching for other scraps, before disappearing back into the trees.

The hunters were used to the proximity of wolves. They even saw a spiritual link with these animals, who were also eking out an existence in these sparse forests on the edge of the true tundra. But this winter, the wolves were more of a constant presence than ever before. They were in the camp every night. In previous years, they had occasionally come near during the hours of daylight – never within the circle of tipis, but close enough. Perhaps they were driven by hunger. Perhaps these wolves had been becoming bolder over years, even generations. Mostly, the humans tolerated them. But stones, bones and sticks were thrown at the wolves if they came too near.

It was at the end of that long, hard winter – surely longer and harder even than the one before – that one wolf, a youngster, came right into the centre of the camp. A girl of about seven was sitting on a log, mending her arrows, and the wolf came very close to her. The girl stopped what she was doing. She laid down the arrows, rested her hands on her knees and looked down at the trodden, compacted snow. The wolf padded a few steps closer. The girl glanced up and down again. Then the wolf came right up to her. She felt the wolf's warm breath on her skin. Then the wolf licked her hand and momentarily sat back on its haunches. The girl looked up and into the blue eyes of the young wolf. An astonishing moment of connection. And then the wolf leapt up, spun round and bounded away, back into the taiga, back into the shadows.

The wolves seemed to be tracking the people that summer, as they in turn tracked the huge herd of reindeer, migrating in stages across the landscape. The snow melted and gave way to vast expanses of grassland. The reindeer would graze and move on. The people were always just one step behind, striking their camp each time the herd began to shift, setting up again when they were settled. Usually, the wolves would melt away in summertime, as hunting became more profitable than scavenging from human hunters. But these wolves – or at least some of them – had somehow found themselves drawn to the side of the humans, even joining in with the hunts – and profiting from the fallen prey.

It was a nervous, fragile alliance. The wolves were wary of the humans, and the humans of the wolves. There were stories of these predators snatching babies from camps, although no one seemed to have experienced this first-hand. There were tales of hunters bringing down a deer, only for wolves to claim the carcass, driving the human hunters away. The older members of the tribe were suspicious and cautious. But there was no doubt that the wolves had improved the success of the hunts. They could help to separate a reindeer or a horse from the herd, sometimes even bringing the animal down before the hunters got near enough to throw their spears. The wolves would flush out smaller game, too. The hunters rarely came home empty-handed. And so there was less hunger – especially during the tough winter months. More wolves ventured into the camp during daylight, and they didn't seem aggressive. After a few more winters and summers, parents would even let their children play with the friendly wolf pups, tumbling and play-fighting in the space between the tents. Some wolves started sleeping close to the camp. It was clear that this pack had affiliated itself with the humans. When the tents were dismantled and wrapped up and the people moved on, the wolves moved with them.

Who domesticated whom? Had the wolves chosen the people or the people chosen the wolves? However it began, this alliance would change the fortunes of the humans and it would change the form and behaviour of their canine companions. After just a few generations, the friendliest wolves had started to wag their tails. They were becoming dogs.

This is clearly fiction. But it is fiction based on scientific facts that we can now be very sure of. Our modern dogs, in all their wonderful variety, are the descendants of wolves. Not foxes, jackals, coyotes or even wild dogs. Wolves. European grey wolves, to be precise. Our modern dogs share over 99.5 per cent of their genetic sequences with these grey wolves.

What drew wolves to our side? Archaeologists in the past have suggested that it could have started with the advent of farming. The lure of livestock – easy pickings for opportunistic predators – would have been hard to resist. But the earliest evidence of farming – marking the beginning of a new age for humans, the Neolithic – goes back some 12,000 years to the Middle East. Dog skeletons have been found at archaeological sites much older than this. Of all the animals and plants that have been changed by coming into close contact with humans, forming alliances with us, the dog seems to be our most ancient ally: the first people to have dogs were not farmers, but Ice Age hunter-gatherers. But just how far back into our prehistoric past can we trace that alliance? And where, how and why did it happen?

Deep in the icy past

The traditional story of the domestication of dogs saw this process taking place around 15,000 years ago, at the tail-end of the last Ice Age. This was the time when the ice sheets were retreating northwards, when trees and shrubs, humans and other animals began to colonise the higher latitudes of Europe and Asia once again. The tundra greened, rivers ran full and the sea level rose, as warmth and life returned to the icy north. The ice sheets that had gripped North America from coast to coast also began to retreat, and groups of humans migrated from the vast, continent-like Beringia into the New World.

There's plenty of definitive evidence of domestic dogs from 14,000 years ago onwards: bones which are clearly those of dogs, not wolves, turn up in archaeological sites across Europe, Asia and North America. Yet there's a possibility that these are relatively late examples. At the beginning of the twenty-first century, as geneticists started to team up with archaeologists to probe questions about the origin of domesticated species, a suggestion emerged: that the domestication of dogs could have begun much earlier, even tens of thousands of years earlier, than previously thought.

Geneticists began to approach the question of dog origins by looking at patterns of differences in dog mitochondrial DNA to reconstruct a 'family tree' for this small package of genes. The results could be interpreted in different ways – the reconstructed family tree was compatible with two, completely distinct models of dog origins. One suggested that dogs arose from multiple origins, around 15,000 years ago. The other fitted with an early, single origin of most dogs – going back 40,000 years. The discrepancy in timing between the models is large – the possible dates are not only separated by thousands of years, but by the peak of the last Ice Age, which climaxed some 20,000 years ago.

Mitochondrial DNA is just one strand, and actually just a tiny part, of the genetic legacy carried inside the cells of an organism. There's much more information to be found in the chromosomes – the packages of DNA contained in the cell's nucleus. There are thirty-seven genes in the mitochondrial genome, compared with some 20,000 in the nuclear genomes (of both dogs and humans). When geneticists moved on to the nuclear DNA of dogs, an earlier date started to look most likely. The first draft genome – the genetic sequence contained in all of the chromosomes – of the domestic dog was published in a paper in *Nature* in 2005. The domestic dog was clearly most closely related to the European grey wolf. The authors (of whom there were – incredibly – over two hundred) had not only worked on a thorough sequence of the dog genome, but had made a start on mapping variation amongst different breeds of dog, looking at where single letters in the DNA sequence varied – at more than 2.5 million positions in the genome. The analysis revealed genetic bottlenecks linked to individual breeds – in other words, the dogs' DNA showed how each breed had started off with a handful of individuals, taking in just a fraction of the genetic variation that had existed across the species as a whole. Each breed represented just a small sample of that variation. Those bottlenecks, linked to the origin of different dog breeds, are really quite recent, probably happening around thirty to ninety generations ago. Assuming an average generation time of three years, that translates into just 90 to 270 years ago. In addition to these more recent genetic bottlenecks, the DNA of modern dogs also held traces of a much more ancient bottleneck: one that was presumed to result from the original domestication of some grey wolves – into dogs. The geneticists estimated that this bottleneck occurred around 9,000 generations ago – around 27,000 years before the present.

This potentially early date for domestication prompted archaeologists and palaeontologists to wonder if they'd been missing something, and one group of researchers set out to examine that possibility. They looked at nine skulls of large canids – animals that could have been either dogs or wolves – from sites in Belgium, Ukraine and Russia, dating to between around 10,000 and 36,000 years ago. They didn't make any assumptions about whether these skulls did in fact represent wolves or domestic dogs. Instead, they made careful measurements and then compared the data from the ancient skulls with a large sample of more recent canid skulls, including obvious examples of dogs and wolves. Five of these ancient skulls appeared to be wolf. One was impossible to pin down. Three were closer to dog than to wolf. Compared with wolves, these canids had shorter, broader snouts, and slightly wider braincases. One of these ancient dog skulls was very old indeed. It was from Goyet Cave in Belgium, which has proven to be a treasure trove of Ice Age artefacts, including shell necklaces and a bone harpoon, as well as bones from mammoth, lynx, red deer, cave lion and cave bear. The cave had clearly been used by humans and animals for thousands – perhaps even tens of thousands – of years. But it was possible to pin a precise date on the putative dog skull, using radiocarbon dating. It was around 36,000 years old – the oldest known dog in the world.

What was particularly interesting about Goyet was that this early dog had a skull shape that was quite distinct from that of a wolf. The palaeontologists who carried out this study argued that this distinct 'dogginess' suggested that the process of domestication – or, at least, some of the physical changes associated with it – may have been very rapid. And once skull shape had changed – from wolf-like to dog-like – it stayed that way for thousands of years.

And yet this is a *single* example of what looks like an early dog, dating from before the peak of the last Ice Age. It's so surprisingly early that it seems sensible to consider the possibility that Goyet is an aberration of some kind. Even if the dating can be trusted, couldn't this just be a weird-looking wolf? Goyet, however, was soon joined by another, apparently very early dog. In 2011, just two years after the publication of the analysis including Goyet, a group of Russian researchers published evidence of what looked very much like another ancient dog, this time from the Altai Mountains of Siberia.

The Siberian skull was found in Razboinichya (Bandit's) Cave – a limestone cavern tucked away in the north-western corner of the Altai Mountains. Excavations starting in the late 1970s and continuing until 1991 turned up thousands of bones, buried in a layer of red-brown sediment deep inside this cave. Amongst the bones were those of ibex, hyena and hare – and a single, dog-like skull. No stone tools were found in the cave, but some flecks of charcoal suggested that ancient people had also visited it during the Ice Age.

In the initial analysis, a bear bone from the fossil layer in Razboinichya Cave was radiocarbon dated to around 15,000 years ago – the late Ice Age. All the other bones were assumed to be of a similar age. So that dog skull could have been boxed up and swiftly forgotten about, languishing on a dusty shelf in a university or museum storeroom – yet another example of a dog from the tail-end of the Ice Age, when the world was warming up again.

But Russian scientists decided the skull deserved more careful scrutiny. Firstly, was it really a dog? The Razboinichya skull – which quickly acquired the nickname of 'Razbo' – was measured and compared with the skulls of ancient European wolves, modern European and North American wolves, and the crania of much more recent dogs, from Greenland, from around a thousand years ago. Those Greenlandic dogs were of a large but 'unimproved' type – they hadn't gone through the genetic mill of extreme selective breeding that would produce all the weird and wonderful variety seen in modern dog breeds. Razbo was a tricky beast to pin down. Like Goyet, its snout was relatively short and broad – a dog-like characteristic. But it had a hooked coronoid process – the projection of bone on the upper mandible where an important chewing muscle, temporalis, attaches – that was more wolf-like. The length of the upper carnassial tooth – a slicing type of tooth, useful for shearing through muscle and sinew – fell within the range for wolves. But this tooth was relatively short compared with other teeth in Razbo's mouth: it was shorter than two molars stacked together – and that's a more dog-like characteristic. The lower carnassial tooth was smaller than that seen in modern wolves, but on the other hand, it fitted comfortably into the range for prehistoric wolves. The teeth were less crowded in the jaw than might be expected for a dog. Despite the short snout, then, Razbo's teeth looked more wolf-like than dog-like. Yet Razbo's skull measurements told another story – the skull shape was closer to the Greenland dogs than anything else.

Of course this was always going to be tricky. Early dogs are only just not-wolves. And while some features of anatomy and behaviour do arrive in package form, often because they depend on just a few genes, most traits will appear in a gradual, piecemeal fashion. The transformation occurs over generations: pieces of the mosaic change, little by little, until the picture is a new one. This is why Goyet was quite remarkable – two distinct changes to skull shape, in the form of a broader snout and a wider braincase – do seem to have appeared very quickly in early dogs. But we shouldn't be alarmed at the discrepancy between skull shape and teeth in Razbo.

With a skull shape like that of a Greenlandic dog from a thousand years ago, but slicing teeth more like wolves, the Russian scientists concluded that Razbo could well have been an incipient dog – one of the earliest examples of this particular experiment in domestication. But still, a 15,000-year-old incipient dog isn't much to bark about. There are plenty of those knocking around. It was the new dating of the skull – direct dating, using bone samples from Razbo itself, and carried out in three separate labs, in Tucson, Oxford and Groningen – that caused a stir. The skull turned out to be around 33,000 years old. Goyet was no longer alone.

Case closed, then: both bones and genes seemed to be pointing to an early date of domestication, around 30,000 years ago, give or take. Rather than being anything to do with the onset of agriculture (at its earliest, some 11,000 years ago in Eurasia) or even the changing environment and society as the Ice Age released its grip (some 15,000 years ago) – it looked like humans' best friend had much earlier origins: way back in the Palaeolithic, before the peak of the last Ice Age, before anyone lived in villages, towns or cities. When we were all still nomads, hunters and gatherers. Long before our ancestors settled permanently in the landscape.

But, unfortunately, the origins of the domestic dog were also far from settled. In 2014, another team of geneticists weighed in to the debate. Various researchers had argued for origins of dog domestication occurring in Europe, East Asia or the Middle East. So the geneticists wanted to look more carefully at the geographic origin of dogs – and to probe the question of whether a single origin or multiple origins was most likely. They sequenced the genomes of three wolves – from Europe, the Middle East and East Asia – as well as an Australian dingo, a basenji (descended from hunting dogs in western Africa) and a golden jackal. The researchers found plenty of evidence of interbreeding between different canid groups, which somewhat confused the issue. Several dog breeds contain traces of quite recent interbreeding with wolves – freely roaming village dogs, for example, probably had fairly regular contact with wild wolves. However, the geneticists were able to sift through the DNA data, looking past these more recent interbreeding events and searching for clues about the earliest dogs, hidden in the genes of their latest descendants. The genetic evidence pointed to dogs having had a single origin of domestication – and they estimated that this took place between 11,000 and 16,000 years ago. This still suggested that the domestication of dogs wasn't linked to the advent of farming, as some researchers had previously suggested. Yet on the other hand, this later date was well after the peak of the last Ice Age, leaving Goyet and Razbo stranded on the other side, deep in time.

But then, these Ice Age dogs had always been controversial. Some researchers had called into question the canine credentials of these animals – they seemed so out of step with the rest of the archaeological evidence. The physical differences between these contentious canids and wolves are, admittedly, quite subtle, and doubts were cast on the methods used to analyse and interpret the skulls. The size of the Goyet canid was considered problematic. With such a large skull, it must have had a large body, too – and domesticated animals are generally smaller than their wild counterparts. So perhaps, some researchers argued, it was just another, now-extinct, variety of wolf, rather than a dog. Or, if Goyet and Razbo really were early dogs, they were probably dead-ends – blips, failed experiments in domestication. The bulk of the archaeological evidence still pointed to the true ancestors of our modern dogs being domesticated much later, after the peak of the last Ice Age. A later date would also go some way to explaining the extinction of Ice Age megafauna, such as woolly mammoth and woolly rhinoceros – hunted to extinction, perhaps, as humans teamed up with their deadly canine companions. The objections to the dogginess of the Goyet Cave canid seemed almost too shrill, too indignant: these early 'dogs' just didn't fit with the current edifice of theory; even if they *were* dogs, they were unlikely to represent the ancestors of our modern hounds. Research into canine domestication is fraught with controversy. If you will forgive me, canine palaeontology is a dog-eat-dog world.

Neither the bones nor the DNA were producing a clear-cut answer, though – in early 2015 it looked as though the weight of evidence was building for a late date of domestication, after the peak of the last Ice Age. After all the excitement over Goyet and Razbo, those early 'dog-like' skulls could just be odd-looking wolves – or early dogs whose descendants died out.

But the date of domestication at 11,000–16,000 years ago, inferred from the DNA of living dogs and wolves, depended on a few crucial assumptions about mutation rates and generation times. If the actual mutation rates had been slower, or generation times longer, *that* would push the date earlier – it would have taken longer for the DNA differences seen between modern dogs and wolves to accumulate.

June 2015 saw the publication of a striking new piece of genetic evidence. This time, rather than sifting through the genomes of modern dogs and wolves, looking for clues to their ancestry, the geneticists had gone after ancient DNA. The transatlantic team, with members based in Harvard and Stockholm, worked on a rib discovered on an expedition to the Russian Taimyr Peninsula in 2010. The rib was clearly canid, and it dated to 35,000 years ago.

Sequencing a tiny section of mitochondrial DNA, the researchers were able to identify the species of the animal to whom this bone had belonged – it was from a wolf. The next part of the investigation involved comparing the ancient genome of the Taimyr wolf with genomes of modern wolves and dogs. The degree of difference between the ancient and modern genomes just didn't tally with previously assumed rates of mutation. Applying the standard rates to the genetic difference between modern wolves and the Taimyr wolf suggested that the common ancestor of both of them lived 10,000–14,000 years ago – but that's less than half the actual age of the Taimyr wolf. So mutation rates must have been less than had been previously thought – 40 per cent of the assumed rate, or even slower. Using the new, low rate of mutation, the predicted date of divergence of wolves and dogs moves from 11,000–16,000 years ago to 27,000–40,000 years ago.

The revelations didn't stop there. The geneticists went on to scrutinise particular patterns of variation in the DNA of modern dog breeds – looking at mutations which each involved a single nucleotide 'letter'. These genetic variants are known as single nucleotide polymorphisms, or – more snappily – SNPs (pronounced 'snips'). These single-letter mutations are good indicators of evolutionary history in the genome because they're common – and often inconsequential, so not weeded out by natural selection. Comparing a handful of SNPs (170,000, to be precise) between modern dog breeds and the Taimyr wolf, the geneticists found that some breeds had more wolf in them than others. This suggests that, after the origin of domestic dogs, some populations had interbred with wild wolves. Breeds that had ended up with a bit more wolf in them included the Siberian husky, Greenland sledge dog, the Chinese Shar-Pei and Finnish spitz. The geneticists also looked at the genetic diversity of modern wolves, and found that the split between North American and European grey wolves must have occurred after the Taimyr wolf lineage had peeled away – but presumably before the sea levels rose at the end of the Ice Age, submerging the Bering land bridge that had – during the glaciation when the sea level was low – provided a link between north-east Asia and North America.

So have Goyet and Razbo been saved, then, by the latest genetic research? It seems there's no reason to doubt the existence of domesticated dogs 33,000–36,000 years ago, nor that their descendants could still be with us today. Genetics, though, has thrown a final spanner in the works here. Goyet's mitochondrial DNA is unusual – distinct from that of both wolves and other dogs, ancient and modern. So we're left wondering what Goyet actually was – an early experiment in domestication that led nowhere? Or an unusual, ancient type of grey wolf that no longer exists today? A sophisticated analysis on the 3D skull shape of the Goyet Cave canid, published in 2015, suggests that it's more wolf-like than dog-like after all. And so the argument continues. Razbo, on the other hand, appears to fit nicely into the dog side of the mitochondrial DNA family tree. So it looks as though Razbo really could have been an early dog – he certainly has plenty of close relations alive today, in the form of our current canine companions.

It's just incredible how heated the debate over the origins of dogs has been over the last few years. New techniques and new discoveries seem to have the potential to radically change theories. And the story keeps changing. But with all the progress – from better dating of archaeological finds, to faster DNA sequencing – the real history of the origin of our oldest and closest ally seems to be emerging from the shadows at last. And it's *bound* to be complicated. Just look at how convoluted the human history that we *know about* is. When we approach prehistory – our own or the unwritten histories of other species – we may start off very naively, somehow expecting a simple story that neatly summarises the complexity of interactions over thousands of years. It's no wonder the picture changes as more scientific analyses are carried out and more detail emerges. The work done on the DNA of the Taimyr wolf and its cousins, ancient and modern, shows just how tortuous tracing the roots of domestication can be.

Having pushed the origin of dogs back into the Ice Age, the next question that emerges is – where were dogs domesticated? And was there a single, discrete area where domestication first began, then spread – or multiple times and places where wild wolves became dogs? This might be impossible to pin down – the domestication of dogs may have started 40,000 years before the present, and interbreeding with wolves continued long after that, and can still happen today. But, armed with the latest genetic techniques, which allow us to unlock secrets from genomes, ancient and modern, we can at least try.

Finding the homeland of dogs

The debate over the date of domestication has rumbled on, but pinpointing the area where dogs were first domesticated is no less fraught with contention. On the one hand, the genetic results are unequivocal: dogs are clearly domesticated grey wolves. But the grey wolf has a huge range – right across most of Europe, Asia and North America today, and its geographic range was even wider in the prehistoric past. So where within the huge territory of grey wolves did the alliance with humans first take off? We can quickly eliminate North America – humans arrived too late, after the last glacial maximum, for the original transformation of wolf into dog to have occurred there. Analysis of wolf and dog genomes provides further evidence that dogs must have evolved from wolves in Eurasia. The family tree of canine genomes reveals an early branching event when North American and Eurasian wolves diverged away from each other, and a later divergence – of Eurasian wolves and dogs. Across the Eurasian range of grey wolves, it's been very much up for grabs: Europe, the Middle East and East Asia have all been put forward as the original homeland of our canine allies.

Geneticists – and by now, you won't be surprised at this – have argued and argued over this particular question. Early analysis of mitochondrial DNA pointed to a possible – single – origin in East Asia. This seemed to be supported by a peculiar shape of part of the mandible that was shared by both Chinese wolves and modern dogs. Genome-wide analyses also seemed to support a single origin, but were less clear about the site of domestication for a while, as wolves from all over Eurasia seemed to be equally related to our modern dogs. Further work on mitochondrial DNA of living dogs from across the world then appeared to settle the question. It revealed what appeared to be a clear connection between all modern dogs and the ancient dogs and wolves of Europe. This seemed to match the archaeology. The bones of ancient dogs have been discovered in East Asia and the Middle East, but the earliest of them dates to just 13,000 years ago – whereas there are prehistoric dogs from Europe and Siberia ranging from 15,000 right back to over 30,000 years ago. The original ancestors of dogs were, most likely, Pleistocene – Ice Age – wolves from Europe.

In 2016, new evidence came to light. Firstly, there was a careful analysis of the part of the mandible that had been thought to indicate a link between Tibetan wolves (*Canis lupus chanco*) and modern dogs, apparently supporting an Asian origin. The coronoid process, where temporalis attaches, was a similar shape in both Tibetan wolves and modern dogs: this large bony projection was unusually hook-like and leaning backward. But a wider study showed that only 80 per cent of Tibetan wolves and 20 per cent of dog mandibles showed this particular trait. It was just too variable and inconsistent to be used to infer an Asian origin of dogs. And then, just as this morphological argument for an East Asian origin of dogs fell through, in 2016, a new genetic study appeared on the scene to liven things up again.

This time the geneticists had surpassed themselves, with a very thorough sequencing of the genome of a 5,000-year-old dog, from the famous Neolithic site of Newgrange in Ireland. They also sequenced mitochondrial DNA from fifty-nine other ancient dogs. They compared all this genetic data with existing data from modern dogs, including eighty full genomes and another 605 sets of SNPs. Firstly, the Neolithic Newgrange dog looked similar in its genes to modern free-living dogs – it hadn't been moulded by the highly selective breeding that would eventually lead to all our modern breeds. And although its DNA suggested that it could digest starch better than wolves, it couldn't do it as well as modern dogs.

However, it was the patterns of variation – or rather, the breaks in that variation – that really caught the researchers' eyes. One modern breed, the Saarloos, stood out from the rest – a little twig on its own, isolated from the rest of the canine family tree. This was not so surprising, as the breed was created in the 1930s by crossing German shepherds with wolves – it's a true hybrid. But there was another deep split in the DNA, driving a wedge between dogs from East Asia and those from Europe and the Middle East. The genome of the Neolithic Newgrange beast clustered, or matched up best, with the western Eurasian dogs. But mitochondrial DNA revealed something else – most of the ancient European dogs possessed different genetic signatures compared with modern European dogs. The geneticists suggested that the ancient dogs of Europe must have been largely replaced by a later wave of incomers from the east.

Hot on the heels of that study, another one reported the results of whole genome analyses on not just one, but two Neolithic dogs – this time from Germany. One dated to the beginning of the German Neolithic, 7,000 years ago (5000 BCE), and the other to the end, around 4,700 years ago (2700 BCE). The early Neolithic dog's genome was very similar to that of the Newgrange dog from Ireland. Yet there were also clear genetic connections, stretching across the millennia, to the late Neolithic dog – and to modern European dogs. There was no sign here of a major population replacement. But there was an intriguing additional signal of ancestry in the later German dog, suggesting that *some* interbreeding had taken place with dogs arriving from further east. This could be a canine echo of a major human migration, westwards from the steppe country to the north of the Black Sea, which saw the Yamnaya culture spreading across Europe. The Yamnaya people were horse-riding nomads, who buried their dead along with pottery beakers and animal offerings, under great mounds of earth. It looks like they may have brought their dogs with them, too – but they blended with European dogs, rather than replacing them. The disappearance of the Newgrange dog's mitochondrial DNA lineage – just one tiny part of its genetic make-up – doesn't have to indicate population replacement. These disappearances, the pruning of particular genetic lineages, happen all the time.

But going back beyond Newgrange, to the origin of domestication itself, what is the meaning of that ancient east–west split in dog ancestry? There are two possibilities. It could be that dogs originated once, then spread out and populations became effectively separated, drifting apart genetically and creating the deep rift. Or there could have been two separate origins of modern dogs, from genetically distinct populations of wolves, one somewhere in west Eurasia, and the other somewhere in east Eurasia. Answering this question turns on the timing of the split – and the date of domestication. The genome sequencing of the two Neolithic German dogs helps to pin down those crucial events in time. Added to the existing data, geneticists came up with a date of divergence between dogs and wolves at 37,000 to 42,000 years ago. The east–west divergence then occurred between 18,000 and 24,000 years ago, after domestication. This means that a single origin – followed by a split – is most likely. What's still up for grabs at the moment, though, is precisely *where* domestication first took place. The only way to settle that question will be to analyse *more* ancient DNA – from even earlier dogs, right back into the Ice Age. At the moment, though, the jury is out. Ancient mitochondrial DNA and archaeological evidence seem to suggest that a European origin is most likely – but genome-wide data from modern and early dogs reveals a hotspot of diversity in East Asia, suggesting that dogs have existed there longer than anywhere else.

This isn't the last word on the origin of dogs, clearly. But it's extraordinary to think how much we have learnt in just the last five years. The early pathfinding of genetics showed us the slender routes laid out by the unfurling, maternal lineages of mitochondrial DNA. The latest techniques, sequencing entire genomes, allow us to see the whole genetic landscape. Questions whose answers have eluded us before are now answerable. The next few years will see our view of the past expanding even more. We already know that dogs were domesticated – most likely somewhere in Europe, when our ancestors were nomadic hunter-gatherers. Soon, we may have a better idea of exactly where that alliance was first formed.

But *how* did domestication of dogs take place – and just how intentional was it? We're so used to thinking of domestication of animals and plants as an idea that occurred to our ancestors some 11,000 years ago as part of the so-called 'Neolithic Revolution', when our forebears gave up their primitive hunter-gatherer lifestyles and settled down to farm, taking control of themselves and their environment and laying the foundation of civilisation itself. There are lots of things wrong with this simplistic view – not least that domestication is a gradual process that is probably much less deliberate, from the human perspective, than we have tended to presume.

First contact

We can only imagine how Ice Age hunter-gatherers and grey wolves teamed up. It probably happened – or nearly happened – many times, in many separate places. There may have been occasions when a tenuous alliance formed, only to break up again. History doesn't run along a railroad, heading for a destination. It meanders, branches off, and often reaches dead ends (and we can only recognise those dead ends retrospectively). But eventually – as we know with the powerful benefit of hindsight fuelled by science – at least one of these alliances prospered and became cemented so that the ongoing partnership of humans and their canine companions was ensured.

What we don't really know is who chose whom. Our instinct may be to suppose that our human ancestors, surely supreme masters of their own destiny, chose the wolves and enslaved them, deliberately moulding them into dogs over generations. In reality, conscious intent may have had very little to do with the transformation of certain wolves into a domesticated species. It may have all started as a gentle form of symbiosis, a loose partnership based on mutual benefit, something more like the story conjured at the opening of this chapter. Perhaps it was even the wolves who drove the process. You don't need to imagine them having some sort of cunning, canine masterplan. By hanging around humans more and more, even if just picking over middens for scraps of food, the wolves may have unconsciously trained the humans into accepting them – first as neighbours, and then as companions.

A successful alliance between our two species must have depended on a predisposition on both sides – a mutual willingness. Both humans and dogs are social animals, but it must go further than that; after all, there are plenty of social animals that we haven't teamed up with. Meerkats, monkeys, mice – none of them have ended up domesticated in the way that dogs have. It seemed possible to me that there was something else, something special about wolf behaviour, that could have paved the way for them forming a bond with humans. In order to find out what that might have been, I needed to get close to some wolves.

High on the ridge above the floodplain of the Severn River, a small pack of wolves roams through ancient woodland. There are just five wolves in the pack, all brothers. Two of them are three years old; three are four years old. They are European grey wolves – slender, compact and long-legged. They are more colourful than their name suggests, with russet flanks, black peppering over their lower backs. Their tails are black at the base and at the tip. Their jaws and cheeks are white. Their pointed black ears are fringed with black fur.

The wolves regularly patrol their territory, loping with a springy, trotting gait along woodland tracks, leaping over fallen trees with flowing, effortless ease. When startled, they run faster, breaking into a canter, but then they will settle, finding a clearing to lie in. When it rains, they find shelter in the undergrowth. They eat meat – from horses, cattle, rabbit and even chickens. But they've never hunted anything larger than a magpie. They don't have to, because the humans looking after them provide them with all the meat they need. This is the captive wolf pack that lives at the Wild Place – a rural enclave of Bristol Zoo, out in the wilderness of South Gloucestershire.

I visited the wolves, staying safely on the outside of their enclosure, with one of their keepers, Zoe Greenhill. She knew the wolves very well, working closely with them every day, and was trying to get them used to being transferred to a smaller enclosure where vet checks could be carried out when needed. That was the limit of the training, though; there was no intention to tame these wolves. And although they'd grown accustomed to Zoe's company, they were still wary of humans in general and easily startled by sudden movements or loud noises. They were also twitchy about new objects in their enclosure; Zoe told me that it had taken quite some time for them to get used to a few newly planted fir trees. I wondered if this group – a small pack of young animals – was particularly nervous, but Wild Place animal manager Will Walker told me that all the wolves he'd ever met had been similarly cautious and shy.

'I've worked with three different packs of wolves in captivity, and I've never experienced any wolves that actively approach you and are confident around you,' he said. 'We work in the enclosures with them – always two of us at a time, in case anything does go wrong – but the wolves always stay away, at the far end of the enclosure. They're so nervous of us – sometimes they even regurgitate their food, before running away.'

'There's surely a conundrum here then,' I suggested. 'If wolves are naturally that cautious around people, how did they ever come close enough to end up becoming domesticated?'

'Well, they are nervous – and if you confront them, they turn and run in the opposite direction. But you *can* play with them. If you turn your back on them, and skip around and hide behind trees, on the other side of the enclosure, they all come running, tails up in the air – and they seem really confident. But if you turn to face them – they're gone again. They're certainly inquisitive animals – they check out what we're doing – but they're not at all bold.'

Of course, it's perfectly possible that wolves have only become this cautious around humans relatively recently – though even people with spears rather than guns would have posed a serious threat to them in the deep past. Caution was surely a good survival instinct. But there was something else which could make wolves overcome their nervousness.

Will told me how the wolves would also follow the keepers when they carried out their morning checks. As the keepers walked around the fence line, the wolves would trot a few steps behind them on the other side. Curiosity must surely have been what first drew wolves to humans. Nevertheless, while the hunter-gatherers were highly mobile, moving on all the time, that inquisitiveness could only ever have led to brief, sporadic encounters – the opportunity simply wasn't there to develop an enduring alliance.

This is where changes in the environment may have played a significant role. Some 30,000 years ago in the Altai, the environment was increasingly conducive to human hunter-gatherer communities becoming settled in the landscape. They were still nomads, but they may have stayed in one place for months at a stretch before moving on. Once people began to settle down a bit more, there would have been time for the relationship with wild wolves to develop. Undoubtedly, the meat brought back by human hunters – and the leftover carcasses – would have been a strong attraction. Eventually, then, curiosity and hunger drove the wolves closer and closer to humans, despite their natural cautiousness. And perhaps the nervousness even acted in their favour. Wolves are large, ferocious-looking animals – formidable predators. But if they looked nervous, rather than too bold, perhaps people would have been less frightened, and more tolerant, of them. From cautious contact to tolerance to partnership – gradually, the alliance between the two very different packs, humans and European grey wolves, grew stronger.

At the point when some wolves started to hang out with humans, their future changed and *they* changed. Wolves who were nervous but friendly would be tolerated. Wolves who were more erratic, perhaps even aggressive, would be driven away or worse. Humans were exerting an evolutionary pressure on the wolves close to them – and the impact of them choosing the friendliest, least aggressive animals would go much further than influencing just that particular facet of their behaviour.

Friendly foxes and mysterious laws

In 1959, the Russian scientist Dmitry Belyaev decided to test out how selective breeding – focusing on specific behaviours – could transform animals over time. He believed that there were fundamental characteristics that were key to dog domestication, that natural tameness would have been positively selected in any wolf cubs, while aggressive tendencies would have been ruthlessly rooted out. He embarked on what has become a famous experiment in domestication, with another species fairly closely related to wolves: silver foxes, *Vulpes vulpes*. Selecting the tamest foxes in each generation and breeding those foxes together, he and his team found that tameness spread quickly through the population. After six generations of highly selective breeding, 2 per cent of the population were extremely tame. After ten generations, that was up to 18 per cent. After thirty generations, half of the foxes were very tame.

By 2006, just about all of the foxes in the ongoing experiment were very friendly towards humans – just like domesticated dogs.

But it wasn't just the foxes' behaviour that had changed. While some of them were still silver, others had turned red. That's still a standard *Vulpes vulpes* colour – not so surprising. Some, however, had turned white with black markings – the so-called 'Georgian White' variety – a complete novelty, never seen in the wild. In fact, the domesticated Georgian White silver fox looks uncannily like a tiny, fox-shaped sheepdog. Some foxes developed a brown mottled colouring, over a silver-white background. Some had floppy ears. And there were changes to their skeletal structure as well, with shorter legs, muzzles and a widened skull appearing. There were changes to reproductive physiology too: wild foxes only mate once a year, but the tame vixens were coming into heat twice, each year. The tame foxes also reached sexual maturity more quickly than their wild counterparts.

Along with the specific attributes of friendliness towards humans, and a lack of aggression, that had specifically been selected for in the experiment, the tame foxes showed other familiar types of behaviour. They held their tails up in the air, and they wagged them. They whined and whimpered to get attention. They sniffed and licked their keepers. They paid attention to human gestures and direction of gaze. By selecting for tameness, the Russian fox-breeding scientists had ended up with a host of other characteristics that seemed to have just come along for the ride but which were also undeniably dog-like.

This fox-breeding experiment shows just how quickly the friendliest and least aggressive wolves from thousands of years ago may have become tamer and tamer with each generation. The hunter-gatherers didn't need to have been selectively breeding like the Russian scientists, following their strict protocol of only allowing the friendliest 10 per cent of their foxes to breed in any generation. The wolf ancestors of dogs could have been self-selecting to some extent – only the friendliest ones would have tolerated living in close proximity with humans. Wolf packs are families, all closely related to each other. If one tended to be tolerant and even friendly to humans, it's likely that others in the pack would have shared the same genes and behavioural tendencies. So it's possible that a whole pack, or even most of a pack, could have formed an alliance. Tame wolves would have been able to form an attachment with humans and to start following human social cues such as gestures and glances. Dogs will make eye contact with humans in a way that a wolf would never do. And dogs have evolved to understand human signals in what can seem to be an uncanny way. Having owned a half-trained Border terrier who rarely did anything I wanted him to do, I was recently astonished at the ability of a springer spaniel to understand my cues. I went for a walk with this spaniel, called Linny, on the shore of Loch Long in Scotland. I threw an old ball for her and it went bouncing off amongst seaweedy rocks. Linny didn't watch it closely enough, and looked at me for some assistance. I shouted, 'Over there, Linny!' and pointed, imagining that I would be clambering over the rocks to retrieve the ball, but she followed the line of my point – perfectly – and found the damp ball in a crevice. I was just as pleased as she was, as she bounded back up the beach to drop it once more at my feet. Linny not only recognised that my pointing finger was a referential cue, she knew what it meant, and how to follow this direction to the damp and smelly prize. She was clearly bred from a long line of dogs who had not simply learnt to attend to human cues, but to follow them to an astonishing degree. Springer spaniels are gun dogs, bred for flushing out game and retrieving the quarry. A soggy ball could stand in for a dead duck. Linny was still delighted to bring it to me. Our modern breeds are relatively new inventions – most are the result of highly selective breeding over just a couple of centuries. But although this uncanny ability to understand human gestures has become finely honed in spaniels, the basis of this behaviour is likely to have emerged a very long time ago. The earliest domesticated dogs probably understood human signals – just as Belyaev's foxes do today.

It seems that domestic dogs – and those domestic foxes – have developed a whole suite of behaviours, as well as anatomical and physiological traits, which seem quite distinct from those of their wild forebears. Yet some of these traits are not entirely novel. I was amazed to learn from Will Walker that wolves occasionally wag their tails and he'd even heard them barking.

'But I've only ever heard them do that as an alarm,' he told me. 'There's an electric fence around the enclosure, and when we first let them in, they were inquisitive and checked it out – touched it – and barked. It just sounded like there was a big dog in there. It was the first time I'd ever heard a wolf bark – but it was a very clear bark. Along with wagging their tails when they're happy – and all the other traits you see in a dog – it's all there.'

This seemed to make a lot of sense; after all, dogs are just domesticated wolves. Many of the traits we associate with dogs haven't just appeared out of nowhere, they were elements of behaviour that were already there in their wolf ancestors. Those traits certainly didn't feature as prominently in the behavioural repertoire of wolves, but they existed. As wolves were domesticated, certain elements of existing behaviour were selected for or promoted and became more common, while others were selected against and pushed out.

Over time, the relationship between tame wolf and human was changing. This wasn't just about two species living side by side and tolerating each other. This was symbiosis; the beginning of a beautiful friendship. Humans were no longer just a source of free food, when wolves could get close enough to the camp. Wolves were no longer simply tolerated, they were encouraged: they clearly had something to offer in return for food. That benefit could well have included companionship, for both adults and children. This is rarely mentioned in theories of domestication, perhaps because it seems too frivolous or fluffy, but I think it's hard to imagine that it didn't play a role. And surely some wolf cubs would have been adopted. Given how much my own kids clamour for a puppy, it's not inconceivable that some Ice Age parent would have given in to this pressure.

But companionship, and amusement for children, certainly wouldn't have been the only benefit of having tame wolves around. Having been a very occasional alarm call amongst wild wolves, a loud bark may have been important in the developing symbiotic relationship between humans and wolves. Maybe those earliest dogs made themselves useful by running with the hunters – helping to track, hunt and even bring back prey. Once farming began, dogs could fulfil a crucial role by protecting livestock from predators such as bears, hyenas – and wolves. But long before that, back in the Ice Age, having tame wolves that could help to protect your camp, and that could sound an alarm by barking, would surely have been very useful indeed.

Barking and wagging tails, then, are not true novelties. We don't need to invoke some new genetic mutations to explain these traits in dogs, because they already existed in wolves. But even if we can explain away some of the differences between dogs and wolves in this way, there still seems to be too much of a gulf between some of the characteristics of dogs and wolves, or between the traits of wild silver foxes and those of the experimental domestic variety, to be biologically feasible. In fact, the same puzzle also exists when we look at the here and now – at the amount of difference within living dogs today. Their variability is just astonishing: from chihuahuas to chow-chows, Dalmatians to dingoes, the range far exceeds what you see in any wild species.

Darwin was intrigued by the amazing variety of domesticated dogs. He suggested that the diversity could have been drawn from a number of different wild canid species, but we now know that dogs came from a single wild species: the grey wolf, *Canis lupus*. In a way, that leaves us with an even bigger question about where all the variety amongst modern dogs comes from. In speculating about the generation of diversity, Darwin thought that a lot of variation could be explained by a multiplicity of environmental factors somehow affecting fertilisation or the development of the embryo. Darwin knew that some characteristics were inherited, but he didn't know *how* they were inherited. And he was very open to the idea that these environmental factors – nurture, if you will – played an important role.

In the early twentieth century, the work of the nineteenth-century monk and scientist Gregor Mendel – who made great inroads into understanding how traits were inherited – was rediscovered, and formed the foundations for the emerging science of genetics. Combined with the observations of naturalists, and with Darwin's mechanism of natural selection, genetics helped to explain how evolution worked. The fusion of these separate branches of biology was described in 1942 by Julian Huxley – the grandson of Darwin's great supporter, Thomas Henry Huxley – in his book *Evolution: The Modern Synthesis*. But the Synthesis had had a difficult birth.

Huxley described how, at the end of the nineteenth century, Darwinism had dug itself into a rut, becoming overly theoretical and overly adaptationist. Every single character of an organism had to be described as an adaptation, wrought by natural selection. Darwinism had become something approaching natural theology – just that natural selection, not a divinity, was cast in the role of designer. At the same time, new biological disciplines had appeared, including genetics, the study of heredity. Experimental genetics and embryology seemed at odds with classical Darwinism.

'Zoologists who clung to Darwinian views,' wrote Huxley, 'were looked down on by the devotees of the newer disciplines, whether cytology or genetics [developmental mechanics], or comparative physiology, as old-fashioned theorists.' But gradually, between the 1920s and 1940s, ideas started to converge – and to make more sense as parts of a whole:

The opposing factions became reconciled as the younger branches of biology achieved a synthesis with each other and with the classical disciplines: and the reconciliation converged upon a Darwinian centre … Biology in the last twenty years, after a period in which new disciplines were taken up in turn and worked out in comparative isolation, has become a more unified science … As one chief result, there has been a rebirth of Darwinism.

The ideas represented by the Modern Synthesis continue to underpin modern evolutionary biology today. We know that the gradual changes which occur within species are essentially due to random genetic mutations. Selection – whether natural or artificial – then acts on those mutations in a non-random way, favouring ones that are advantageous, weeding out any that are not. Still, the variability of domestic species, and of dogs in particular, seems to be just *too* extreme to be explained only through the accumulation of genetic changes over time – by the simple interaction between random, new mutations in genes and selective breeding. Selection can quickly lead to the spread of advantageous genes (and traits) through a population, but it can't speed up the underlying rate of mutation.

Belyaev certainly thought that something else – not just mutations in DNA – must have been responsible for all the changes he saw happening in his increasingly tame foxes. It wasn't just the speed of the changes that required explanation, but the striking similarity between domesticated silver foxes and dogs. It was impossible to believe that all those traits in the foxes – from tail-wagging to floppy ears – arose through new mutations and that the similarity with dogs was coincidental. It seemed unlikely that each individual trait had appeared in an entirely piecemeal fashion. Instead, it seemed more probable that one or two fundamental genetic changes were having widespread effects – that genes were working in a hierarchy, with some controlling others.

And just possessing a particular gene is only the start of the story – genes can be switched on and off. Belyaev hypothesised that the genes which controlled behavioural variation also played an important regulatory role – affecting a cascade of other genes, switching them on or off – during development. The Russian scientists who inherited Belyaev's experiment have suggested that the genes in question may be involved with the hormone cortisol, which mediates the body's stress response, and the neurotransmitter serotonin. The domesticated foxes had very low levels of cortisol in their blood, and higher levels of serotonin in their brains. Low cortisol levels have also been shown in other domestic animals, while high serotonin levels are associated with an inhibition of aggression. But what's really important here is the possible effect of these two biological signals on a developing fox-cub fetus.

The Russian scientists suggested that maternal cortisol and serotonin could influence how many other genes are expressed, both during embryonic development and even after birth while the cubs are still suckling. By choosing foxes which were particularly tame, the Russian scientists may have been selecting individuals who had certain variants of a few, key genes relating to stress tolerance and diminished aggression. This meant that the next generation of foxes could be exposed to unusual patterns of stress hormones, in the womb, which could in turn affect the way that genes were being switched on and off in the developing fox fetus – in a way that didn't normally happen in the wild. The programme of embryonic development, which had settled into a fairly stable state under natural selection, certainly seemed to be shaken up in some way, producing a surprising degree of variety amongst the increasingly domesticated silver foxes. The researchers suggested that just a few genetic variants could have widespread effects, introducing a range of coat colours as well as oddities such as drooping ears and even curly tails. Other researchers have suggested that changes to thyroid hormones – and related genes – could have similar widespread effects on stress response, tameness, body size and coat colour. So selective breeding focusing on one particular trait, likely to be linked with genes relating to stress tolerance and tameness, could quickly affect a whole host of other characteristics.

We've just begun to identify some of the genes that could be involved in creating such a range of effects and to understand how this happens at a molecular level. Geneticists have started to comb through dog genomes to look for particular regions, particular stretches of DNA, that look as though they have been subject to selection. It's a tricky thing to do. The complicated population history of domesticated dogs, which includes migrations, the extinction of some populations, interbreeding in some places and genetic isolation in others, makes it a difficult task. Nevertheless, there are regions of the genome that stand out, and eight out of the top twenty identified regions contain genes that have important neurological functions. One of them is already known to have effects on both social behaviour *and* pigmentation. It's called ASIP, the Agouti Signalling Protein gene. The protein it encodes switches the pigment-producing cells known as melanocytes in hair follicles to producing a paler version of melanin – essentially it controls how darker and paler fur develops in different areas. For good measure, ASIP affects fat metabolism as well – and has also been shown, in mice, to influence aggressiveness. This one gene illustrates quite beautifully how selectively breeding animals which exhibit a certain type of social behaviour could lead to incidental changes in colouring and metabolism. But some traits which end up being inherited together may be influenced by separate genes – that, crucially, lie in close proximity to each other on a chromosome. Strong positive selection for a particular trait, and a particular gene, will often mean that neighbouring genes come along for the ride.

The idea that different traits can be somehow linked together, and inherited together, has been around for a long time and even predates genetics. It's called pleiotropy (Greek for *many characters*) and the term was coined in the early nineteenth century. In the *Origin*, Darwin wrote, '... if man goes on selecting, and thus augmenting, any peculiarity, he will almost certainly unconsciously modify other parts of the structure, owing to the mysterious laws of the correlation of growth.' These laws are much less mysterious today – we know that various traits are linked through genetics and development. We understand the precise basis of the correlations in at least some cases – such as that of the Agouti Signalling Protein and its widespread effects in the body. Combined with the idea of destabilising selection, where artificial breeding must be regularly bringing particular sets of genes together, pleiotropy goes a long way towards explaining why dogs are so much more variable than wolves whilst being, on the face of it, genetically very similar. New genetic mutations can have widespread – pleiotropic – effects, influencing a whole range of characters. And in some cases, you probably don't even need a brand new mutation to spice things up, you just need to be combining particular sets of genes that don't usually get pressed together quite so consistently in the wild. In this way, the developmental programme is destabilised – throwing up new and interesting varieties in the process. It seems very likely that even amongst early dogs, and long before any of the modern breeds emerged, there was plenty of variability – just as there is in the experimental, domesticated silver foxes.

The initial domestication of wolves into dogs could have been – if not as fast as the transformation of wild silver foxes into domesticated ones, over just fifty years – still relatively swift. The new theories about the underlying molecular mechanism of the change reveal pleiotropy at almost every turn. The cascading and destabilising effects of specific genetic variants, picked out initially for their influence on docility and tolerance, have the potential to create widespread and potentially very rapid changes to anatomy, physiology and other aspects of behaviour. What seems like a difficult and improbable transition – from wild to domesticated – suddenly appears to be a much easier, and even likely, development. Perhaps there were many, many instances of wolves becoming dogs, or almost-dogs – even if we can only find genetic traces of just one or two of these experiments developing into lineages that survive to the present day.

The big chill of the last glacial maximum, peaking between 21,000 and 17,000 years ago, put animals right across Eurasia under pressure. Ice sheets descended over Europe, and Siberia became incredibly cold and dry. Many lineages went extinct. Sometimes entire species succumbed. It wouldn't be surprising if more than a few canine domestication experiments were curtailed by this environmental catastrophe. In the run-up to the glacial maximum, free food at the margins of hunter-gatherer camps could have made all the difference to some packs of wolves.

Everyone felt the chill; humans too. And even if some lineages of ancient dogs went extinct, experts have argued that having dogs may have been a crucial survival advantage for human hunter-gatherers at the peak of the last Ice Age. Could this even explain why modern humans, though hard hit, made it through the last glacial maximum, while Neanderthals did not? It's a neat and enticing explanation, but this always makes me nervous. I suspect it's way too simple. History is complex, and while we can suggest hypotheses, we have to be wary when we can't even begin to test them. Nevertheless, there seems no reason to doubt that dogs would have helped the survival and success of some tribes of hunter-gatherers.

After the big chill, fossil evidence of domestic dogs starts to appear all over Eurasia. By 8,000 years ago they're found at sites stretching from western Europe to eastern Asia. As we've seen, the latest genetic data from ancient and modern dogs points to a single origin, so it's extremely unlikely that all these Holocene dogs were independently domesticated from local wolf populations. Instead, dogs must have arrived with migrating humans, or were acquired from elsewhere by local human populations.

Prehistoric dogs were still fairly wolf-like, judging from their skeletons at least. But there was probably already quite a bit of variety in coat colour, tail curliness and ear floppiness, if those Russian foxes are anything to go on. At the 8,000-year-old site of Svaerdborg in Denmark, archaeologists have found evidence for three types of dogs of different sizes. So it looks as though, even this early, there was some divergence into what might perhaps be seen as proto-breeds. Maybe our prehistoric ancestors were already trying to breed dogs that had particular skills: dogs for guarding and shepherding, dogs good at following scents, or even at pulling sledges.

A breed apart

After the origin and expansion of agriculture, dogs became even more widespread. And just as human diets were changing, it seems that dogs' diets were too. Early dogs were eating meaty diets – although, one study suggests, perhaps different meat from their wild wolf cousins. Analysis of bones from the 30,000-year-old site of Předmostí in the Czech Republic has shown that the canids thought to be Palaeolithic dogs were eating meat from reindeer and muskox, whereas the wolves were eating horse and mammoth meat. Once agriculture started, the menu of food available from humans would have changed. There must have been rich pickings for village dogs hanging around the rubbish dumps of newly sedentary human communities.

Most modern dogs have multiple copies of the amylase gene, which encodes the enzyme for digesting starch. The more copies a dog possesses of this gene, the more amylase it produces in its pancreas – extremely useful if you're finding food in the village midden or eating scraps from the table. Over time, dogs' diets became less carnivorous, and more omnivorous – more like the diets of their human allies. But the number of copies of the amylase gene varies considerably amongst modern dogs. Most of the variation in amylase gene numbers comes down to the breed. There could be a few reasons for this. Having established that this variation wasn't just down to chance, researchers wondered if it could be linked to phylogeny – to the 'family history' of breeds. But that didn't seem to be the case. They also considered whether interbreeding with wolves might have reduced the copy number of the amylase gene in some breeds, but again that didn't appear to adequately explain the pattern. The explanation that's left standing is that amylase copy number reflects differences in ancient dog diets.

Studies of isotopes of carbon and nitrogen from samples of ancient dog bones have revealed clues to ancient diets – showing just how variable those diets were. We know, for instance, that around 9,000 years ago in China, millet made up 65–90 per cent of what dogs were eating. Whereas 3,000 years ago, on the coast of Korea, dogs were devouring marine mammals and fish. In various places, dogs were being exposed to different dietary challenges. Over time, their genetic make-up changed accordingly.

This type of change to the genome – boosting the numbers of a particular gene – happens because of mistakes during meiosis, the special type of cell division that makes egg or sperm (which contain a single set of chromosomes, as opposed to the double set contained in all other cells of the body). During meiosis, the chromosomes pair up, and then, in each pair, swap DNA with each other. Mistakes that happen during this 'crossing over' can result in duplications of a gene on one chromosome. Once that occurs, it actually increases the chance of a similar mistake happening in the next generation, again at meiosis, when eggs or sperm are made. Two copies of a gene on one chromosome, and one on another, make mis-pairing and gene duplication more likely. So this error can end up multiplying copies of a particular gene – and if that change is beneficial natural selection won't weed out those mistakes but favour them.

Dogs seem to be split into two groups – those with very low numbers of the amylase gene, and those with many copies. The modern dogs with the lowest number – just two copies, like wolves – tend to be from breeds such as the Siberian husky, Greenland sledge dog and the Australian dingo. Dogs with high copy numbers map rather neatly on to agrarian areas of the globe – where humans were farming in prehistory. The saluki, which originated in the Middle East – where agriculture first got off the ground – has a phenomenal twenty-nine copies. But this change wasn't immediate – Neolithic dogs don't show the major expansion in amylase genes that their later descendants, living alongside farmers, would evolve.

It's in the Neolithic, when humans start to farm, that dogs also start to spread beyond Eurasia for the first time. And they track the spread of farming. Dogs appear in sub-Saharan Africa after the beginning of the Neolithic there, 5,600 years ago, and take another 4,000 years to reach South Africa. Dogs appear in archaeological sites in Mexico around 5,000 years ago, coinciding with the first farmers there, but only reach the southernmost tip of South America 4,000 years later. Studies of mitochondrial DNA suggested that all those early American dog lineages were completely replaced, following the European colonisation of the Americas. But the latest genome-wide studies tell a different story: European dogs – arriving with colonisers in just the last 500 years – mixed with the indigenous New World dogs.

The modern breeds that we know so well take much longer to arrive. They are *very* recent inventions. Dog genes reflect this history. There are signs of two prominent genetic bottlenecks amongst the ancestors of dogs: one at the origin of domestication, and another when modern breeds emerged, in just the last 200 years. Breeders began to focus closely on promoting particular traits, producing dogs that were wonderfully obedient, providing invaluable help with hunting and herding. But the malleability of characteristics under selective breeding became an allure in itself, and so dogs were also bred with specific shapes, sizes, colours and textures. The morphological variety amongst modern dog breeds exceeds that in the whole of the rest of the family Canidae, which includes foxes and jackals as well as wolves and dogs.

There are nearly 400 breeds of dogs today, and most of them – in all their wonderful diversity – have really only been around since the nineteenth century. This is when the strict breeding needed to create and conserve the kinds of strains recognised by kennel clubs really got going. The breeds that appear to be most ancient, with the most deep-rooted lineages on the dog family tree, are actually found in places where dogs only arrived relatively recently. Dogs arrived in the islands of south-east Asia 3,500 years ago and in South Africa around 1,400 years ago and yet these areas are home to a number of 'genetically ancient' breeds: basenjis, New Guinea singing dogs and dingoes. This pattern shows that these lineages have been isolated for longer than most other breeds. The deep roots don't mean that their lineages were the first to branch off, but rather that out on the periphery they have stayed the most genetically distinct.

Analysis of the genomes of various dog breeds has been used to build a very detailed family tree. Within that family tree, there are twenty-three clusters or clades, each one containing a set of branches which represent a group of closely related breeds. European terriers, for instance, form one clade; basset, fox and otter hounds, together with dachshunds and beagles, form another. Spaniels, retrievers and setters are also a closely related cluster. Strict control of breeding has kept these clades largely separate – but a few breeds contain DNA from two or more clades, revealing how different dogs with particular traits have been recently crossed to create new types. For instance, while the pug dog has genetic connections with other Asian toy breeds, as expected, it's also part of the tight cluster containing European toy dogs. This suggests that pugs were exported from Asia, and then deliberately crossed with European dogs to create new, diminutive breeds. Although the genetic data reflects the creation of strictly separate breeds in the last 200 years, it's also clear that these breeds weren't drawn from a homogeneous population – selection for distinct traits had already separated dogs into types which were suited to particular functions, and those older distinctions form the basis for those twenty-three clades in the dog family tree.

Lots of breeds with supposedly ancient roots, however, turn out to be recent recreations. Wolfhounds were, as their name suggests, used to hunt their wild cousins – very successfully. By 1786 there were no wolves left in Ireland, and so no need for wolfhounds. By 1840, the Irish wolfhound had also gone extinct. But then a Scotsman living in Gloucestershire, Captain George Augustus Graham, resurrected the 'Irish wolfhound' by breeding what he thought was a wolfhound of some kind with Scottish deerhounds. Today's population of Irish wolfhounds comes from a very small group of ancestors so that, like many breeds, they are *inbred*. And while this helps to maintain the characteristics of the breed, it also increases the risk of particular diseases with a strong genetic component. Around 40 per cent of Irish wolfhounds suffer from some form of heart disease, and 20 per cent from epilepsy. They're not the only pedigree with problems. Many dog breeds plummeted to near-extinction in the twentieth century, during the world wars, and were resurrected by outbreeding with other types of dog. Very strict breeding since then has produced extremely inbred populations, with little genetic diversity within breeds and an increased risk of diseases – ranging from heart disease and epilepsy, to blindness and particular cancers. Specific breeds are predisposed to certain afflictions: Dalmatians have a high risk of deafness; Labradors often suffer from hip problems; cocker spaniels are prone to developing cataracts.

Breeds may be relatively reproductively isolated now, but their genes tell us that there was once plenty of gene flow between breeds or proto-breeds. Breeds from separate countries share characteristics and genes which show that they must have interbred in the past. The Mexican hairless dog and the Chinese crested dog share hairlessness and missing teeth and in both breeds these traits are caused by precisely the same mutation in a single gene. The odds against this gene mutating in exactly the same way in two different dog populations are infinitesimally small. Instead, these shared traits and shared genetic signature speak of common ancestry. Dachshunds, corgis and basset hounds all have very short legs. Together with sixteen other dog breeds, they all have exactly the same genetic signature associated with this form of dwarfism – the insertion of an extra gene. It's most likely that this insertion happened just once, in early dogs, long before any of the modern short-legged breeds appeared.

Genetic research provides us with this astonishing opportunity to understand the evolutionary history of dogs, from the pleiotropic exuberance of variety produced by selecting tameness, right through to the selection of peculiar features, suited to very particular tasks, in our modern breeds. We can see how certain mutations, and the traits associated with them, popped up amongst early dogs, and were later

– much later – promoted and propagated by selective breeding to create the modern breeds we know today. With inbreeding producing problems with increased risk of disease, geneticists are also working to understand the basis of particularly prevalent diseases and it may be possible to reduce that risk by even more careful selective breeding, and judicious outcrosses, underpinned by genotyping.

Some breeds have been outcrossed beyond the bounds of domestic dogs. Such extreme outcrossing was the basis of the Saarloos wolfdog, created in 1935 by breeding together a male German shepherd with a female European wolf. The Dutch breeder, Leendert Saarloos, was hoping to create a more ferocious and formidable working dog, but he ended up with a meek and cautious beast. Saarloos wolfdogs make good family pets, and have been used as guide dogs and rescue dogs. Another breed – the Czech wolfdog – was also created by crossing a German shepherd with a wolf, this time in 1955 in Czechoslovakia. The Czech wolfdog, originally bred for military service, has also been used for search and rescue and is increasingly popular as a pet. Will Walker owns a Czech wolfdog, called Storm. 'She's just as friendly as any other dog. She loves every dog and every human she sees,' he told me. She makes an excellent guard dog, too. 'She'll bark at anything – she's very keen to defend me and my house.' 'You're like those early hunter-gatherers with wolves protecting their camps!' I commented.

But the growing popularity of wolfdogs – spurred on by the appearance of these impressive animals in *Game of Thrones* – is balanced against increasing concern about their suitability as household pets. There's an important distinction to be made between animals bred from recent hybridisation and those established breeds such as the Saarloos and Czech wolfdog, which are genetically much more 'dog' than 'wolf'. Some breeders of wolf-dog hybrids, however, offer animals which are advertised as the product of much more recent crosses, which raises concerns about the potential for wild, unpredictable behaviour.

Hybrid wolf-dogs have attacked and killed a number of children in the US, and are banned outright in some states. In others, wolf-dog hybrids are legal, as long as the hybridisation happened at least five generations ago. In the UK, a first- or second-generation wolf-dog hybrid is considered risky enough to be regulated by the Dangerous Wild Animals Act – the same law governing owning a lion or tiger. It seems odd that breeders would exaggerate the wolf content of their puppies – but wildness is part of the cachet of these animals. With buyers seeking 'high-content' and 'wild looks', and willing to part with £5,000 to feel more like Jon Snow, wolf-dog hybrids are big business. It's difficult to know just how 'wolfy' the product of a cross is, several generations down the line. The first-generation animals will be 50:50 in their genes, but after that, the shuffling of DNA that happens as eggs and sperm are made introduces messiness – second-generation wolf-dogs could have up to 75 per cent wolf genes in their genome, or as little as 25 per cent. There's also the possibility that some purported 'wolf-dog hybrids' are nothing of the sort, and are just cross-breeds of German shepherds, huskies and malamutes – which already look fairly wolf-like – to create animals which appear even more like wolves. The 'wolfiness' of a wolf-dog hybrid, a few generations after hybridisation, is impossible to pin down without genotyping. And even with that genetic measure of wolfiness, it's difficult to know how this would relate to the potential behaviour of an individual animal.

There are also concerns about wolf-dog hybrids on the other side as dog genes make their way into the genomes of wild wolves. Genetic studies have shown that 25 per cent of Eurasian wolf genomes contain dog ancestry. This is problematic from a conservation perspective – could an injection of domestic dog genes into wild, grey wolves cause problems for *Canis lupus*? Wolf populations have declined in Europe, under pressure from both hunting and the fragmentation of habitats. But hybridisation could also supply beneficial genes and traits. North American wolves got their black coat colour by interbreeding with dogs centuries, if not millennia, ago. Most hybridisation appears to occur through free-ranging male dogs mating with female wolves, but one recent study showed up dog mitochondrial DNA in two Latvian wolf-dog hybrids. Mitochondrial DNA is exclusively inherited from the mother, so the only way that this DNA could have ended up in wolf genomes is by female dogs having mated with male wolves. Once dog genes have entered wolf populations, it's very difficult to remove them. Some hybrids look a bit like dogs, but many look exactly like wild wolves. So experts have advised that the best way of reducing the impact of hybridisation is to reduce the number of free-ranging dogs. Once they mate with wild wolves, it's too late.

Hybridisation raises all sorts of questions. There are biological questions about the integrity of species, and about just how much interbreeding occurs across our once sacrosanct species boundaries. If there's plenty of interbreeding, with fertile offspring, does this mean our species boundaries are too narrow? These are widely debated questions right now. But in fact, taxonomists, the people who make it their business to name and circumscribe species, have never been quite as rigid as textbooks may have led us to believe. Species are simply snapshots of evolutionary lineages – diverging (and sometimes converging). They are defined by being diagnosably different from the nearest cousins on the tree of life. But sometimes they are defined for human convenience – especially when it comes to conferring separate species names on domesticates and their wild ancestors.

The potential for hybridisation also leads to ethical questions about the 'contamination' of wild species with genes from domesticated species. Having created domesticated species, we're now keen to preserve any surviving, closely related wild ones. But does this invoke an idea of the purity of species that just doesn't really exist in the real world? That's a challenging question, and one that will only become more pressing as our own population grows, and the species we've become allied with burgeon alongside us. It's such a conundrum. The species that have become our allies have secured their future, by becoming companionable, useful, even indispensable to us. But together, we represent a threat to whatever wildness remains.

It seems that the safest way for humans and wolves to co-exist on the planet is to avoid each other. Our ancestors once tolerated wild wolves – long enough to domesticate them. Wolves may be naturally much more shy around humans now than they were in the past. Wolves were changed by becoming domesticated dogs, in so many ways, but the wild wolves may have changed as well. Persecution and hunting of wild wolves probably exerted a selection pressure of its own – the most successful wolves are likely to have been the ones that stayed away from humans. Wolves that are more fearful, and that avoid us, may be products of human-mediated selection – as much as dogs are.

The genetics of grey wolves and dogs suggests that the wolf lineage which gave rise to dogs is now extinct. Times were tough around the last glacial maximum, so that's certainly possible. But there's another way of looking at the family tree – that particular lineage of wolves is not extinct at all; in fact, it's the most populous branch of the wolf family tree: dogs. Genetically speaking, dogs *are* grey wolves. Most researchers simply subsume them within the grey wolf species, *Canis lupus* – not a separate species, the previously recognised *Canis familiaris*, but a subspecies: *Canis lupus familiaris*.

So that terrier, that spaniel, that retriever that you know so well . . . it's a wolf at heart. But a much friendlier one – even more tail-wagging, hand-licking, and altogether less dangerous – than its wild cousins.

2

WHEAT

Triticum

History ... celebrates the battlefields whereon we meet our death, but scorns to speak of the ploughed fields whereby we thrive; it knows the names of kings' bastards but cannot tell us the origin of wheat. That is the way of human folly.

Jean-Henri Casimir Fabre, nineteenth-century
French botanist

A ghost in the ground

Eight thousand years ago, a seed fell on to the fertile ground, somewhere near the coast, in north-west Europe. It had travelled far. Not windblown, not even carried on the beak or in the guts of a bird. But on a boat. It had been part of a precious cargo, but it was so small it fell to the ground in a clearing in the forest, and no one noticed.

The seed started to germinate. It sprouted and its long leaves grew up. But the weeds around it were stronger. The interloper never managed to produce its own seeds. It died back. Nevertheless, its ghost remained in the earth. Even after the saprotrophic fungi and bacteria had done their best to pull apart every last shred of its being, a few molecules of that exotic plant survived. And with every year that passed, that layer of soil was buried deeper, as the forest floor built up. Then trees disappeared, to be replaced by sedge and reeds. They grew and died and half-rotted down. The sea level was rising, and the reedbeds were replaced with samphire and seablite. The rising tide brought in fine sediment to create a layer of mud over the peat. For a time, this new mudflat was only inundated at the highest, spring tides. Then twice a day. Then it was submerged and even the seablite couldn't hang on any more. The sea level rose, and the waves rolled in. But the molecular ghost of the ancient, exotic plant still lingered in the deep peaty sediments – buried under metres of marine clay – at the bottom of the Solent.

A lobster makes an archaeological discovery

In 1999, a lobster living on the seabed, close to Bouldnor – just to the east of Yarmouth – on the north coast of the Isle of Wight, made an astonishing discovery. The lobster had been digging its burrow into the base of a submerged sea cliff, excavating sand and cobbles out of the bank.

Two divers spotted the lobster, and the excavated ditch that it had made, leading to its burrow. The furrow ran alongside an ancient, fallen oak tree. And in it, the divers found stones that the lobster had pushed out of its burrow. The divers were maritime archaeologists, interested in the well-preserved submerged forest at Bouldnor Cliff. They picked up the stones that the lobster had excavated – and saw that they were man-made: worked flints. These weren't the first stone tools that the archaeologists had discovered in this area, but the others had eroded out of the sediments and had been shifted by currents. The lobster's flints looked as though they had only moved a short distance – and the divers suspected that the original context of these artefacts may well be in the cliff itself, just where the lobster had chosen to make its home.

The submarine archaeologists set to work – diving for an hour at a time, surveying and excavating the area at the base of Bouldnor Cliff. Despite poor visibility and the strong currents, they found an astonishing wealth of archaeological material, and started to build up a picture of the environment, when it was dry land. They found relics of an ancient forest – pine, oak, elm and hazel. Alder was there too, a tree that likes to grow with its feet in the water, perhaps along the banks of an ancient river. And in the sandy sediments that must have once been the banks of that river, the archaeologists found evidence of human activity: plenty of flints, some of them burnt, together with charcoal and charred hazelnut shells, and the oldest piece of string in Britain. Radiocarbon dating revealed that the site had been occupied around 6000 BCE. Nearby, the divers found evidence of a pit containing burnt layers, and a mound of timber that might, just might, represent a raised platform that once supported a Mesolithic house. There was also plenty of worked timber, with the marks of ancient tools still clear. These timbers included a large piece of split oak – possibly part of a log-boat – and a wooden post, still standing upright in the ancient sediments. The preservation was fantastic. It became clear that, after the site had been abandoned in antiquity, peat must have grown over it quickly, sealing the archaeology *in situ*. And there it all lay, just waiting for that fortunate lobster to come along and discover it, 8,000 years later.

The underwater excavations at Bouldnor Cliff went on from 2000 to 2012. Analysis of all the material will take many more years. There's just so much – from an archaeological and a palaeo-environmental perspective – for a diverse team of researchers to get their teeth stuck into. Along with all the obviously archaeological material that the divers brought up from the seabed – the chipped flints, the fragments of charcoal, and the carbonised hazelnut shells – they brought up mud. Lots of it. These samples of sediment would undoubtedly contain more tiny clues to the prehistoric environment at Bouldnor Cliff – perhaps minute rodent bones, small pieces of plant, and even pollen – which would be yielded up by sieving and microscopy. But in

2013, another team of researchers contacted the Isle of Wight archaeologists. They wanted the mud, but what they hoped to find wouldn't be visible under even the most powerful microscope. They were after molecules. Long, stringy molecules, rich with information. They were after DNA.

The geneticists approached the Solent mud with open minds. They didn't start with preconceptions about what they might find, and then try to find it (or not). They looked at samples from the layer that included the hazelnut shells, and applied a technique known as 'shotgun sequencing' – as indiscriminate as the name suggests. It sounds like the antithesis of hypothesis-driven research – the gold standard that all good scientists should be striving for: *the* Scientific Method. But there isn't just one 'Scientific Method'. Sometimes the best way of starting out to understand something better is simply just to ask – what's out there? And then you collect data and try to make sense of it. Arguably there is still a hypothesis even in such broad approaches – which directs which data are to be collected – but there's not an experiment as such, just good *looking*. Much of genomics works in this way – amassing large amounts of data and looking for patterns. In this case, the hypothesis was expansive: 'ancient DNA from contemporary organisms will be found in the sample'. And although it may be heretical to say it, I think it's when you keep hypotheses as broad as possible, when you break away from any preconceptions and expectations, that you can have the best chance of finding something truly novel and exciting.

The geneticists working on the Bouldnor Cliff mud pulled out all manner of DNA sequences from things that had lived there, 8,000 years ago (6000 BCE). They found genetic traces of oak, poplar, apple and beech trees, as well as grasses and herbs. *Canis* was there too – either from dogs or wolves, and *Bos* – which must have come from aurochs, the ancient ancestor of cattle. The molecular ghosts of deer, grouse and rodents were also hidden in the sediment. Piece by piece, the geneticists put together the details of the ancient ecosystem of the Solent forest, where the Mesolithic hunter-gatherers had made their camp.

But amongst the fragments of DNA from the seabed, there was something which came as a complete surprise: the unmistakable trace of *Triticum*. Wheat. It shouldn't have been there. This was Britain, pre-farming. The samples of sediment had already been checked for pollen – usually a good indicator of the plants growing at the time. But there was no wheat pollen in the samples. Was it a mistake? It was such an unusual finding, the geneticists had to be absolutely certain that they weren't looking at something else. But the *Triticum* sequence seemed real enough. The team checked carefully to make sure that the signal couldn't have come from some other wheat-like grass that was indigenous to Britain – perhaps lyme-grass, couch grass or wheatgrass. But the ancient DNA was different to all of these. Instead, the closest match was one particular species of wheat: *Triticum monococcum*, or einkorn – 'single corn'. Each of the little spikelets in the ear of this wheat contains a single seed, enclosed in a tough husk. Einkorn was one of the first cereals to be domesticated and cultivated but it wasn't thought to have arrived in Britain until 6,000 years ago (4000 BCE) – a full 2,000 years later than the unmistakable genetic traces of it at Bouldnor Cliff.

So the einkorn buried in the sediments at the bottom of the Solent had travelled far and fast to get there, so long ago. The birthplace of cultivated einkorn was two and half thousand miles away, at the eastern end of the Mediterranean. And the first person to start focusing in on the original homeland of einkorn, and other wheats, was a botanist and geneticist born in Moscow in 1887.

Vavilov's courageous quest

In 1916, the 29-year-old Nikolai Ivanovich Vavilov left St Petersburg to embark on an expedition to Persia – modern-day Iran. He had a particular objective in mind: to track down the origins of some of the world's most important crops.

Vavilov had studied in Britain, under the eminent biologist William Bateson. Through Bateson, Vavilov would have become familiar with Mendel's ideas about inheritance. William Bateson had helped to revitalise and popularise the work of the Augustinian monk Gregor Mendel – including Mendel's famous experiments with pea plants. Mendel had worked out that there must be some kind of 'units of inheritance', influencing whether his peas ended up green or yellow, smooth or wrinkled. He had no idea what those units were – we now know them as genes – but he predicted their existence. Mendel published his 'Principles of Heredity' in German, in 1866. More than forty years later, Bateson translated this seminal work into English, and it was he who came up with a name for the scientific study of inheritance, based on Mendel's observations and theories: 'genetics'.

Vavilov was also familiar with Darwin's theory of evolution by natural selection. While in England, he spent plenty of time poring over the books and notes in the personal library of Darwin – kept at the University of Cambridge, where Darwin's son, Francis, was Professor of Plant Physiology. Vavilov saw for himself how carefully and comprehensively Charles Darwin had studied the works of his predecessors, including the influential German botanist, Alphonse de Candolle, who had explored the origin of domesticated plants in two weighty volumes, published in 1855. Vavilov clearly enjoyed tracing the development of Darwin's ideas in the notes scribbled in the margins, and at the end of these books. He admired Darwin's thorough scholarship, his distillation of ideas, and his clear understanding of biological processes. 'Never, before Darwin, had the idea of variation and the enormous role of selection been advanced with such clarity, definition and substantiation,' he wrote.

Nikolai Vavilov believed that Darwin's ideas were crucial to pinpointing where species – including domesticated ones – had first sprung into being. Darwin's ideas about the geographic origin of species – articulated in *On the Origin of Species* – were essentially very simple. The origin of any species was likely to be the place which still possessed the greatest variation of types within that particular species. This is still a guiding principle in modern studies: the place with the greatest genetic – and phenotypic – diversity today is probably the place where that species has existed longest. It's a useful guide, but runs into problems because, over time, plants and animals don't stay put. But Vavilov believed that variation in closely related wild species could also be an important clue – so he spread his net a little wider, looking at wild relatives as well as the domesticated crops he was interested in.

Vavilov worked as a state botanist – his specific remit involved researching domesticated varieties of plants in order to inform Russian agronomy and plant-breeding. But he was equally intrigued by the historical and archaeological dimensions of his work. He believed that pinpointing the origins of domestic species would also be important to 'explaining the historical destiny of peoples'. He also realised that, in elucidating the origin of domesticated wheat, there would be insights into a crucial moment in human history, when our ancestors moved on from simply gathering wild foods to growing them: when they made that transition from foragers to farmers. Vavilov knew that he was looking for history before history. The earliest domestication of species would have happened long before writing was invented. He wrote: 'The history and origin of human civilisations and agriculture are, no doubt, much older than any ancient documentation in the form of objects, inscriptions and sculpture reveals to us.'

The pursuit of the origin of domesticated species had long been the preserve of archaeologists, historians and linguists – but Vavilov believed that botany and the new science of genetics could make an important contribution. In fact, he was quite disparaging about the nature of the traditional evidence. 'Philologists, archaeologists and historians speak of "wheat", "oats" and "barley",' he wrote in 1924. 'The present state of botanical knowledge demands that the cultivated wheat species be distinguished into 13 species, oats into 6 species, all quite different.'

And he knew that his was no armchair science. He needed to get out there. He needed to understand landscapes and the plants that grew in them. And above all, he needed samples. 'Every single packet of grain,' he wrote, 'every handful of seeds and every bundle of ripe spikelets is of the utmost scientific value.'

Vavilov came back from his Persian expedition with evidence of a huge variety of types of cultivated wheats. He divided wheat species into three groups, each with a different number of chromosomes. Species of soft wheat, including common or bread wheat (*Triticum vulgare*), had twenty-one pairs of chromosomes. Hard wheat, including emmer wheat (*Triticum dicoccoides*), possessed fourteen pairs, and einkorn (*Triticum monococcum*) had just seven pairs of chromosomes. Back in Russia, as few as six or seven varieties of soft wheat were grown. In Persia, Bokhara (in modern-day Uzbekistan) and Afghanistan, Vavilov recorded some sixty distinct varieties. He was clear that south-western Asia must be the homeland of this form of cultivated wheat. The distribution of hard wheats was a little different, with the most variation occurring in the eastern Mediterranean. Einkorn was different again – wild varieties of einkorn were found across Greece and Asia Minor, Syria, Palestine and Mesopotamia. He observed: 'Most likely, the region of Asia Minor [Anatolia] and the areas adjacent to it appear to be the centre of einkorn variation.'

Vavilov believed that these separate centres of domestication for each type of wheat had influenced the characteristics of the various species in ways that were still relevant to him as an agronomist – as someone who was interested in improving crops. Hard wheats, like emmer, originated along Mediterranean coasts, where spring and autumn are wet, and summers dry. They needed moisture to germinate and start growing, but were quite drought-resistant when mature. Vavilov believed that emmer wheat was the earliest domesticated form of wheat – he wrote about it as 'the bread wheat of ancient agricultural peoples'. And he had an intriguing theory about the later origin of einkorn.

When the earliest farmers began growing wheat, they found that certain other plants seemed to enjoy living alongside the sown crop. They had discovered weeds. And some of those weeds would eventually become domesticated themselves. Wild rye and oats were both common as weeds in fields of wheat and barley. Vavilov suggested that rye started to be grown as a crop by allowing the weed to replace the wheat over winter, or on poor soils or in harsh climates – where rye was hardier than the crop that had originally been sown. When Vavilov travelled around Persia, he saw fields of emmer wheat heavily infested with a weedy variety of oats. He suggested that farmers attempting to grow emmer wheat in more northerly latitudes would have found the oats taking over their fields. The farmers were effectively forced to adopt oats as a crop.

Vavilov provided many other examples of plants which he believed had started out as companion weeds, before becoming crops in their own right. Flax started as a weed amongst linseed crops. Garden rocket started as a weed in fields of flax. Vavilov noted that wild carrots commonly appeared as weeds in vineyards in Afghanistan, where, he wrote, 'they practically invited themselves to be cultivated by the local agriculturalists.' Similarly, cultivated vetches, peas and coriander probably originated from weeds in cereal crops. And Vavilov suggested that one of the grassy weeds infesting Anatolian fields of emmer would go on to become an important cereal itself – einkorn.

But back in Russia, Vavilov's ideas were not popular. Darwin's theories and Mendelian genetics were not in vogue in Stalin's Soviet Union. Vavilov himself began to be seen as a threat; a dangerous weed. His student, Trofim Lysenko, whom Vavilov described as 'an angry species', stuck the knife in. Whilst on an expedition to the Ukraine, Vavilov was arrested, and incarcerated in Saratov Prison. He never left the prison, dying of starvation there, in 1943.

The Crescent and the sickle

Following Vavilov's brave and pioneering work on the origin of crops, further botanical and archaeological evidence accumulated, to secure a grand sweep of the Middle East as the 'cradle of agriculture'. Encompassing the land between and around the Rivers Euphrates and Tigris, and stretching across to the valley of the Jordan, this 'Fertile Crescent' has become renowned as the birthplace of the Eurasian Neolithic – one of the earliest places in the world where farming began. This is where the first domesticated wheat, barley, peas, lentils, bitter vetch, chickpeas and flax – all the plants which have become famous as the 'founder crops' of the Eurasian Neolithic – emerged. Recent studies have suggested that broad beans and figs should be added to this list.

Archaeology reveals the presence of very early farming communities, in what is now Turkey and northern Syria, between 11,600 and 10,500 years ago. But there's evidence that people in the Middle East were exploiting wild cereals long before they domesticated these crops. Traces of domesticated cereals – including barley, emmer wheat and einkorn – are often found in shallower, more recent archaeological layers, directly above deeper and older layers containing the traces of their wild counterparts: the first wheat, barley, rye and oats to appear in archaeological contexts are gathered wild cereals.

Thousands of grains of wild barley and oat, dating to between 11,400 and 11,200 years ago, have been discovered at Gilgal in the Jordan Valley. Evidence of wild rye with some early signs of domestication – fatter grains with indications of having been threshed – has been unearthed at Abu Hureyra, on the Euphrates. And in some places, there is intriguing evidence of what the hunter-gatherers were doing with the wild cereals they were gathering.

At sites across the Southern Levant, the existence of small carved-out hollows in rocks has puzzled archaeologists for decades. Some have suggested that these cup-holes could represent the output from ancient masonry competitions. Or that they could symbolise genitalia. (Now, I totally accept that some cultural artefacts may indeed represent such important elements of anatomy – it would be weird if they didn't. But it's very hard not to see the interpretation of any old bump or hole as sexually suggestive as more indicative of the mind of the archaeologist than that of the ancient creator of such an artefact.) In any case, a more prosaic explanation for these particular hollows seems much more likely: that they are mortars used for food preparation – specifically, for grinding cereal grains into flour.

Many of these purported mortars have been discovered at Natufian sites – belonging to a culture which was well established by 12,500 years ago, a good 800 years before the earliest glimmers of the Neolithic in the area. The culture got its name from a cave in the Wadi an-Natuf, in the West Bank, excavated by Dorothy Garrod in the 1920s. The archaeological term for the period in which the Natufian occurs is the Late Epipalaeolithic. This means something like 'peripheral Palaeolithic' – it's a term ripe with the implication and expectation of change. Society and culture was evolving, in a way which is clearly seen in the archaeology, but it's not quite Neolithic yet.

The Natufian culture in the Southern Levant emerged around 14,500 years ago, and brought with it an important shift – from restless wandering to sedentism. The Natufians were still hunter-gatherers, but they were *settled*. They lived in permanent, year-round villages, rather than temporary camps. And by 12,500 years ago, then, these villagers were carving out stone cup-holes which look like mortars. The only cereal with large grains to grow in this area at this time was wild barley. And so one group of archaeologists recently decided to put these rock mortars to the test – how well would they work to grind barley grains into flour?

The experiment was as authentic as the archaeologists could make it. While they may not have been dressed up as ancient Natufians to carry out the test, they made sure that the whole procedure was performed using Natufian-style tools. First, they harvested wild barley with a stone sickle – previous experiments had shown that cutting stalks with reproduction flint sickles produced precisely the same polish seen on archaeological flint tools, interpreted as sickles. Then they gathered up the spikelets into a basket. Then they used a curved stick to thresh the barley, to separate the awns – the long bristles – from the spikelets. Then the spikelets were pounded in a conical mortar, with a wooden pestle, to remove the awn bases and the husks. The chaff was winnowed away by gentle blowing. Then the naked grains were returned to the mortar to grind them into flour, using the wooden pestle in stirring and pounding motions. The archaeologists finished off by using the flour to make a dough which they baked into an unleavened, flat bread, similar to pitta, over the coals of a wood fire. Then they ate their experimental loaf, and presumably, went for a beer.

The archaeologists had used an actual, ancient rock-cut mortar, at the site of Huzuq Musa, for their experiment. There were thirty-one narrow conical mortars at this site, as well as four large threshing floors close by. Based on their experiment, the archaeologists reasoned that the Natufians at Huzuq Musa could easily have processed enough barley for this to have been the staple food for the hundred or so inhabitants, 12,500 years ago. And it was important that the conical mortars seemed to work so well for de-husking the cereal grains. Barley with husks on could have been made into groats, porridge or coarse flour. But de-husked barley can be ground into much finer flour – and there really is only one reason for doing that: to make bread. It's astonishing to think that the ancient inhabitants of Huzuq Musa could have been gathering barley, threshing it and grinding it into flour, and breaking bread with one another, at least a thousand years before anyone started to grow any cereal crops.

The idea that bread had already become a staple of Middle Eastern diets, hundreds of years before the inception of agriculture, makes the Neolithic Revolution easier to understand. In fact, once people had started to gather and process wild grains, I think domestication of those species – not only barley but wheat and other cereals – was almost inevitable. If you come to rely so much on one particular food, then perhaps depending on a harvest of wild grains becomes too risky. Better to grow some yourself. But this suggests our ancestors deliberately set out to cultivate wild plants. It's likely that the beginning of agriculture owed much more to happenstance and serendipity than any carefully laid plans.

It seems that at least some of the changes that mark out domesticated cereals from their wild predecessors may have come about accidentally, or at least, as unintended consequences of human actions. A crucial difference between wild and domesticated cereals lies in the strength of the central spine, or rachis, to which the seeds are attached – forming the ear of wheat. In wild types, the rachis is brittle, and shatters: the individual spikelets that contain the seeds break away from the ear as they ripen, scattering themselves to the wind. An ear of domesticated cereal, on the other hand, remains intact after ripening. The rachis is tough – not at all brittle. This is a characteristic which would be severely disadvantageous to a wild grass – the seeds cannot be lost freely into the wind and scattered. In the wild, it would be a problematic mutation that would be swiftly weeded out by natural selection. But in a crop, the tough rachis becomes an advantage.

If harvesting was left until most ears had ripened, then any with a brittle rachis would already have lost many of their seeds – but the mutant, tough-rachis plants would still be hanging on to all their spikelets. So all of those still-clinging seeds would be carried to the threshing floor – some to be eaten, some to be sown again. And so the proportion of tough-rachis seeds and plants would increase with each generation. It's another example of a certain characteristic almost selecting itself. Farmers did not need to be actively seeking out particular plants that were holding on to all their seeds. All they needed to do was to wait until most of the wheat was ripe, and then the wheat they harvested would be relatively enriched with the tough-rachis type – so the spread of this particular characteristic could well have been an unintended consequence of early farming practices.

In fact it's possible that selection for a tough rachis started to operate even before farming began. Imagine being a hunter-gatherer, bringing back armfuls of wild cereals to your settlement to process them. You'll drop plenty of seeds on the way. But if any of the wheat you've gathered has a tough rachis mutation – those ears will stay intact. When you get back and start threshing, it's inevitable some of those grains will escape, germinate and grow. Did the first fields spring up around the threshing floors, before any sort of cultivation was practised? It's certainly a possibility but, ultimately, tough-rachis wheat would need to be sown. The characteristic may have developed as an unintended consequence of the way in which cereals were being harvested and processed, but once particular strains of wheat had evolved like this, they were trapped in their alliance with humans – the plants could no longer survive without our help. They could only grow on the edges of the threshing floor – or in fields where they were deliberately sown.

The tough-rachis feature spread through ancient wheat, slowly but surely, over some three thousand years, as people started to depend more and more on cereals, and to cultivate them. A few sites in the Levant have turned up a small proportion of non-shattering einkorn or emmer as early as 11,000 years ago. But by 9,000 years ago (7000 BCE), plenty of sites have 100 per cent non-shattering wheat: the trait has clearly become the norm – in the language of genetics, it has 'become fixed' – in populations of ancient domestic crops.

The transformation of wheat, from wild to domesticated, was a protracted process. That slow transformation was accompanied by a similarly slow change in the tools used by the hunter-gatherers-turned-farmers. Gradually, more and more sickles start to appear in archaeological sites. Unlike the more familiar, curved metal blades, the first sickles were made from flint or chert – this is still the Stone Age, after all. They are long blades which would have been fitted in wooden handles (archaeologists know, because just a few have been found like this, intact). The characteristic 'sickle sheen' along their edges shows that they were polished by repeatedly being used to cut the silica-rich stems of grasses. Sickles don't appear out of the blue – they were probably tools that had been used for a long time for cutting reeds and sedges, before they were used for harvesting armfuls of wild cereals. From around 12,000 years ago, sickles become a little more frequent in the archaeological record – mostly in the Levant, the western limb of the Fertile Crescent. Archaeologists interpret this increasing use of sickles as indicating a new dependence on cereals – as it seems unlikely that the people of the Levant started to obsessively cut more reeds.

Around 9,000 years ago, sickles become even more common, right across the Fertile Crescent. But they're not entirely ubiquitous – leading some archaeologists to suggest that the use of sickles may be more of a cultural preference than a total prerequisite for cereal harvesting. This isn't as surprising as it sounds: evidence suggests that hand-plucking wheat and barley – as the Bedul Bedouin in the Valley of Petra still do – may be just as efficient as harvesting with a stone, or even a metal, tool. Perhaps the increasing use of sickles in the Near East between 9,000 and 6,000 years ago had more to do with the cultural identity – a 'badge' of farming – than the efficiency of the harvest itself. Nevertheless, the rising number of sickles seems to be more than just symbolic, reflecting a real increasing dependence on cereals – which initially represent just a small proportion of gathered plants, at a handful of archaeological sites. But by 7000 BCE, most sites where plant remains are preserved show cereals in the majority. And the wheat that was being cut and gathered in wasn't just clinging on to its spikelets – the grains were larger than those of their wild predecessors. Again, something which would be a disadvantage in the wild – seeds too large to be dispersed by the wind – becomes a bonus for the farmer.

Some increase in grain size is seen in wild wheats before the appearance of the tough-rachis feature. And then grains get bigger and bigger over three to four thousand years. Some of the increase in size is definitely down to genetic changes, but a proportion of it is probably environmental – as crops benefit from being grown in prepared soil, having to compete less with weeds, and are even well watered.

A grain of modern, domesticated wheat comprises three important components. There's the plant embryo or germ – this is a seed, after all. Then there's the seed coat (the pericarp and testa) which makes up about 12 per cent of the weight of the grain, and is commonly known as bran. But by far the bulkiest part of the grain is the endosperm – making up 86 per cent of the grain weight. Like the yolk of an egg, the endosperm is there to provide sustenance to the developing wheat embryo. It contains starch – a load of starch – as well as oils and proteins. And it's the endosperm that expanded, disproportionately, as grain size increased – packing more nutrition into each and every grain of wheat. But the embryo *did* increase in size as well – nowhere near as much as the endosperm, but still significantly. And there's one really important characteristic of large-grained cereals when it comes to germination and early growth – their seedlings are much more vigorous than their small-grained counterparts.

It seems reasonable to assume that an increase in grain size would have come about through conscious, deliberate selection of large-grained plants by early farmers. But, once again, this trait may have been selected for quite inadvertently. Early farmers were probably focused on increasing the size and productivity of their fields, rather than the size of individual grains. Larger-grained strains of wheat, with more vigorous seedlings, may simply have had an inbuilt advantage, out-competing smaller-grained varieties. Competition between seedlings – which may not have happened so much with wild, wind-scattered varieties – could have become fierce in densely sown, cultivated fields. Slowly, from summer to summer, the fields would have filled with larger-grained varieties, much to the farmers' delight.

These two important traits – the tough, non-shattering rachis and the larger grain size – didn't develop at the same time in each species. They're clearly not traits which come along as a package, like tameness and coat colour in dogs. They evolved at different rates, and for different reasons. And – rather like the initial steps towards domestication amongst the wolves who started to follow Ice Age hunter-gatherers – it seems that the process may have started off with much less forethought by humans than has often been assumed. But even without specific intent, the actions of humans produced a significant change in these cereals – making them even more productive, almost by accident. As domestication traits spread and became fixed in these plants, they became even more valuable to humans. Wheat became more and more important in ancient diets – and its future as a dietary staple was ensured.

The protracted and complex history of domestication in wheat almost comes across like the plot-line of a romantic novel. Two potential partners – in this case, a *Homo* species and a *Triticum* species – meet. They are thrown together and then could so easily go their separate ways. But the contact has awakened something in each of them. They start to dance with each other. They grow together. Human culture changes to embrace *Triticum*; wheat changes, to be even more attractive to humans.

The partnering up of humans and wheat is a little more complicated, though. For one thing, there's not just a single type of wheat. Modern botany still recognises the three broad groups of wheat identified by Vavilov, characterised by varying numbers of sets of chromosomes. And modern genetics has revealed the complex relationships between them.

Einkorn, both wild and domesticated, belongs to the group with a simple, double set of chromosomes: just seven pairs. It is, in the language of genetics, a diploid organism (like you and me). At some point in the distant past, there was an ancient doubling-up of chromosomes in one lineage. This happens from time to time, essentially as a mistake in cell division. The cell doubles up its chromosomes but then fails to divide in two – and persists as a single cell with double the number of chromosomes. An ancient doubling-up like this created the tetraploid wheats, with fourteen chromosome pairs (or pairs of pairs of seven chromosomes, if you prefer). This happened between 500,000 and 150,000 years ago – a very long time before the Neolithic Revolution – in the ancient, wild ancestors of emmer wheat and durum wheat.

Then a hybridisation event occurred, involving domesticated emmer wheat (tetraploid) and wild goatgrass (diploid), producing a type of wheat with twenty-one pairs of chromosomes – three sets of pairs: a hexaploid plant. This hybridisation is estimated to have happened some time around 10,000 years ago – and it produced *Triticum aestivum*: common wheat or bread wheat.

A doubling-up of chromosomes seems greedy enough. Most organisms can get along perfectly well with two sets of chromosomes. Four sets seems unnecessary. Six sets seems extraordinarily profligate. But many plants exhibit polyploidy – possessing multiple sets of chromosomes – and it doesn't seem to do them any harm. In fact, it can create significant advantages. The presence of extra genes means that, if one gene has been damaged by a mutation, there's another one available to take its place, and fulfil its function. The mutated gene might even end up doing a new and interesting job in the genome. Bringing together genetic material from different sources – as happened when emmer hybridised with goatgrass – can also lead to hybrid vigour, as novel combinations of genes start to work together, even without new mutations. In addition, polyploidy tends to be associated with an increase in cell size in plants, and it can result in larger seeds and better yields too. It's not all rosy, though – being polyploid can be problematic. Reproduction gets a little trickier, with all those sets of chromosomes to sort out. And embryonic development can get confused, sometimes in a lethal way. But, in bread wheat at least, the evolution of hexaploidy certainly seems to have been – on balance – a good thing.

In particular, the productivity of bread wheat was enhanced by a specific genetic mutation which led to an unusual shape of ear. The wild ancestors of this wheat possess flat ears, with spikelets arranged, in a staggered fashion, on each side of the central spine or rachis. But in bread wheat, a single, favoured mutation produced something quite different: a square ear, with densely packed spikelets – the classic shape of an ear of wheat that makes it look so distinct compared with other grasses. It seems likely that the emmer-goatgrass hybrid that we know as bread wheat, *Triticum aestivum*, may have immediately been a productive cereal that the early farmers would have recognised and cultivated.

And so wheats partnered up with humans, forming a bond which would last for millennia and only get stronger over time. But where did it all start? Precisely where in that grand sweep of the Fertile Crescent did each of these wheat crops – einkorn, emmer and bread wheat – originate?

The Middle East has been a Mecca for archaeologists for two centuries, and the geographic origin of the Neolithic founder crops has been at least one of the holy grails being sought. But even with the new discipline of archaeobotany, with its precise approach to individual species – that Vavilov would surely have approved of – the origins have proved somewhat elusive and obscure – until very recently.

Here, there or everywhere

The Fertile Crescent is a vast area which takes in parts of modern-day Israel, Jordan, Lebanon, Syria, Turkey, Iran and Iraq. As we've seen, the seeds of cereals – first wild forms, later replaced by domestic varieties – are found in archaeological sites across this region. It's also a place where the distributions of separate species of wild wheats, barley and rye also overlap. But it's a huge area. Vavilov focused in on each species, carefully recording and sampling varieties of domestic and wild species, and using his data to suggest a homeland for each. For a while, it seemed as though genetics and archaeology were in step.

The great Australian archaeologist Gordon Childe, a leading light at the Institute of Archaeology in London, saw the invention of agriculture as a momentous step-change in human history. In 1923, he coined the term 'Neolithic Revolution'. The transition from hunter-gathering to farming was like a regime change. The old order was overthrown. The New Wave surged across Mesopotamia and the Levant, sweeping all in its path. It all came together beautifully – ideas would ripple out from a centre of creativity; new species would spill out from their centres of domestication. The archaeologists' 'Neolithic Package', including all the founder crops, mapped neatly on to Vavilov's 'Centres of Origin'. The northern arc of the Fertile Crescent seemed to be the focus for the revolution that would change the world. An elite group of proto-farmers in the Near East was brave enough to tame nature – then their burgeoning population spread, taking their new idea with them.

Geneticists moved on from looking at numbers of chromosomes – as Vavilov had done – to decoding the DNA within them. By the 1990s, the technology had advanced to a point where geneticists could look at several equivalent sections of DNA in different plants, and compare their sequences. It was a more powerful technique than just looking at one small area of the genome. And so they looked at wild and cultivated varieties of einkorn and found that the domesticates formed a neat family tree with a single origin. Einkorn seemed to have evolved from one, discrete population. The DNA of domesticated einkorn was closest to that of wild varieties growing in the foothills of the Karacadag Mountains of south-eastern Turkey. Wheat with two sets of chromosomes – including emmer – looked very similar under such analyses. It, too, bore the marks of a single origin – and once again, that seemed likely to have been in Karacadag. Barley also looked like it had come from a single origin, this time in the Jordan Valley. The new, molecular science of genetics had waded into the debate about the origin of domesticated crops – and settled it. The results emanating from these studies which dealt in molecules not middens, which somehow delivered certainty in a way that archaeology never could, and which were revered enough to be published in the most widely read scientific journals, appeared to be decisive.

So it seemed that Vavilov and Childe were right: cereal crops had been domesticated, rapidly, in discrete centres – before spreading out as the craze for farming took hold. The old idea of the Neolithic Revolution was vindicated. There really were core areas, and a single origin, for each domesticated species. It even seemed that a select cultural group in south-eastern Turkey might have come up with an amazing idea which would propel them towards an elite status and allow their population to swell and spread.

It makes for such a great story. If only it were true. But by the early twenty-first century, cracks had begun to appear. Both archaeologists and archaeobotanists were arguing that domestication was likely to have been a protracted and complicated process. Archaeobotanical evidence from the Euphrates Valley suggested, for example, that it took a thousand years for domesticated einkorn to develop the tough rachis – the backbone of the ear of wheat – that would prevent its spikelets shattering off the ear before threshing. This could only be squared with the genetic data if the early domesticates were kept rigidly separate from their wild counterparts, eliminating any possibility of hybridisation, right from the start. But that seemed desperately unlikely.

Archaeologists searching for the roots of agriculture had at various times suggested specific areas within the Fertile Crescent as the likely birthplace of the Neolithic Revolution. Kathleen Kenyon's work at Jericho in the 1950s, digging down into the Neolithic layers there, led to the proposition that agriculture began in the Southern Levant. Other archaeologists favoured the northern and eastern edges of the Fertile Crescent – the hilly flanks of the Taurus and Zagros Mountains. And then the 'Golden Triangle', between the Tigris and Euphrates Rivers and the Taurus Mountains – where wild forms of many of the 'founder crops' overlap with each other – looked like the core area. But as the archaeological evidence accumulated, it seemed to be increasingly pointing to a networked domestication of crops across a much larger region. It also seemed that the early history of agriculture was littered with false starts and dead ends. There was no clear progression. In the Middle East, the Neolithic got going in patchy and sporadic fashion, across a wide area and spanning millennia.

Two lines of evidence, then – the archaeological and the genetic – providing contrasting views of the process of domestication.

Computer simulations suggested that the genetic results might be untrustworthy – the techniques probably couldn't reliably tell apart a crop that had come from a single origin versus one with multiple origins and plenty of interbreeding. But still the idea of a fast, local origin for a great range of domesticates held sway – until dissent began to appear within the genetic camp too. As geneticists spread their sequencing net wider, they also started to uncover more complexity.

The first clue that the single-origin, core area, paradigm might be an artefact of the method, rather than a real, reliable finding, came from more in-depth analyses of barley. Plants have extra packets of DNA – separate from that in chromosomes – inside their chloroplasts (the miniature factories in plant cells that 'do' photosynthesis). Sequencing of particular sections of the chloroplast DNA of barley suggested that this cereal came from at least two separate homelands. The region of the barley chromosome involved specifically with the mutation that produces non-shattering ears bore the same message. More research led to the conclusion that barley had been domesticated, not only in the Jordan Valley, but also in the foothills of the Zagros Mountains. The more detail the geneticists looked for, the more they uncovered. The most recent, genome-wide analyses of barley have revealed that various strains of domesticated barley show genetic connections with nearby strains of wild barley. There seemed to be not just one wild ancestor – but many. One review seems to suggest a wonderfully lyrical source for these new insights: 'Poets and colleagues have recently demonstrated a pattern of genetic diversity in barley that runs entirely counter to the view of a centric origin.'

But it turns out that Ana Poets is a geneticist (although – who knows – she may be a poet as well; after all, it's rare to find a scientist who has no artistic side), and the lead author on a recent paper about shared mutations in domesticated and wild strains of barley. She and her colleagues showed that domesticated barley, far from having a single origin, had a mosaic ancestry – it had been drawn from a wide range of wild strains, each leaving their marks scattered across modern genomes. Apparent connections with wild strains could, of course, be due to much more recent interbreeding – but the team ruled this out. The genetic mosaic of the barley crop had much more ancient roots.

When geneticists looked at emmer wheat more closely, it also turned out to have a more complicated ancestry than initially thought. Domesticated emmer wheat remains had been found from various sites across the Fertile Crescent dating to more than 10,000 years ago. But the earliest genetic analyses showed that the entire domestic species was closest to a discrete population of wild emmer growing in south-eastern Turkey. This seemed to suggest that agriculture emerged in a tiny, core area of the Fertile Crescent, probably around 11,000 years ago. But then the story changed – later studies showed that emmer wheat was also mosaic, closely affiliated with many different wild strains, across a large swathe of the Middle Eastern landscape.

A similar story emerged for einkorn. The original genetic enquiry into its origin suggested a single, discrete focus of domestication. But by 2007, more detailed analyses had started to produce clues to a more complex birth of the domesticated crop: there was no reduction in genetic diversity – no domestication 'bottleneck'. Instead the genetic variation in the crop was drawn from a very wide sample of its wild progenitors, right across the northern arc of the Fertile Crescent.

With the story of barley, emmer and then einkorn evolving in the same way, it seems that multiple, parallel centres of domestication may be the rule rather than the exception for cereal crops. A small 'core area' for domestication, in south-eastern Turkey, simply isn't borne out by the evidence today. Genetics has now converged with archaeology: there were numerous, connected 'centres' of domestication in the Fertile Crescent. This diffuse origin of crops may have been extremely important to the success of domesticated strains, ensuring that adaptations to local habitats passed from wild into domesticated forms. This makes so much sense – local wild species would already be adapted to local conditions. Any farmer taking the seeds of a cereal domesticated in the cool, moist foothills of the Karacadag Mountains, and trying to grow them in the hot, arid plains of Southern Levant, would be unlikely to meet with success.

But some adaptations would have been useful outside the original locality. A particular chunk of genome, which hails from wild barley growing in the Syrian Desert, has been found in a great variety of domestic strains, both in Europe and Asia. This stretch of DNA spread through domestic barley strains, and was preserved – it's likely that it confers an important physiological advantage, such as drought resistance. The sharing of genes between separate populations of early domesticates is clearly evidence of some interbreeding. And it seems that these connections reflect more than just windblown and bird-borne seeds travelling around the region. The human communities of the Near East were well connected: similarities in material culture show that ideas were travelling around. But goods were being exchanged as well – there's evidence of the sought-after volcanic glass, obsidian, passing from one community to another, in a way that we might be brave enough to call trade. It seems entirely reasonable to presume that both knowledge about cultivation and seed corn itself would have also been exchanged between different communities. But even with a 'trade' in seed corn, it's also clear that the primary domesticates grown in various areas across the Near East during this dawn of the Neolithic were local, wild plants – not species brought in from elsewhere.

In case this all seems like ancient history (which it is – and surely, interesting enough), there are implications of these insights into domestication and the genetic basis of particular traits that could be extremely important to us. If the precise effect of that stretch of Syrian-wild-barley DNA could be elucidated, for instance, then this knowledge could be used to improve crops in the future. We shouldn't just view domestication purely as something that happened a long time ago, and which has no relevance to us now. Undoubtedly, there was an intensive period of biological change in crops between 10,000 and 8,000 years ago – including the evolution of large grains and a tough rachis. But domesticated species never *stopped* evolving – and we're still influencing that evolution, perhaps more deliberately than ever before. Vavilov knew that studies into the deep past of cultivated crops would generate useful tools for modern agronomy. Nearly a hundred years later, that's still the case, and the convergence of genetics and archaeobotany is highlighting all sorts of genes – and other areas of the genome – which might be usefully encouraged or even modified. Today's efforts at improving cereals are just the latest steps along a road which started, even before humans began to sow and tend crops, with the gathering, threshing and milling of wild grains; with the baking of bread.

This is all looking good, then. Genetics, archaeology and archaeo-botany have harmonised. We have this coherent story of people seriously exploiting wild grains by 12,500 years ago, and probably even making flatbread from finely ground flour; the emergence of cultivation of cereals from around 11,000 years ago, and the gradual domestication of species, at multiple, connected centres. By 8,000 years ago, most of the wheat and barley being grown across the Near East is non-shattering and large-grained.

There's always more to learn. The current state of knowledge – as I write – certainly won't be the last word on the domestication of wheat. The story is likely to change at least a little, as new evidence is discovered and analysed. But it seems at least unlikely that the whole mountain of evidence that has been gathered in at this point will be completely overturned. It feels as though we have the backbone of the story, and it's unlikely to shatter. We know, well enough for now, the when, where and how of the story of wheat. But we haven't yet cracked – at least in this telling of the story – the *why*.

And this, perhaps, is the most interesting question of all. Because wheat is, at its heart, a grass. A humble grass. Surely not the most obvious foodstuff. Once you've got as far as grinding some grass seeds into a fine flour to make bread, as those ancient Natufians probably did with wild barley, then yes – I can see the attraction. But how do you get to that point? The small seeds of wild grasses seem so unappealing as food. There are surely plenty of other seeds, nuts and fruits which are more tempting: tasty morsels which don't require anywhere near so much hard work to render them edible. What was happening, twelve and half thousand years ago, to make people look at something as humble and unattractive as *grasses* as a source of sustenance? What led our ancestors to depend on such an unlikely food? And why did it happen *then*?

Of temperature and temples

There's a huge lag between the first evidence of wild wheat in archaeological sites – going back some 19,000 years ago – and the earliest evidence of morphologically distinct, domesticated wheats, some 8,000 years later.

At the Syrian site of Abu Hureyra, domesticated cereals gradually replace wild grains between 11,000 and 10,500 years ago. The cultivated species include einkorn, emmer wheat and rye. It's almost impossible to tell which of these species was domesticated first. Radiocarbon dating is extremely accurate, but it always provides a range rather than a specific year. Nevertheless, it's been suggested that einkorn, with its simpler set of seven chromosome pairs, may have been the earliest species of wheat to have been domesticated – rather than being the later weed-turned-domesticate that Vavilov suggested it was.

But why were all these grasses domesticated from the ninth millennium BCE onwards – not earlier, and not later? The timing of domestication suggests that external forces may have been important.

After the peak of the last Ice Age, around 20,000 years ago, the world started to warm up. This was a bad thing for cold-adapted animals and plants, whose habitats were shrinking, but for temperate, warmth-loving species – including us humans – things were suddenly looking up. By 13,000 years ago, the ice sheets of the northern hemispheres had retreated, leaving behind fragments of ancient ice as glaciers, high up in mountain ranges, and covering Greenland and the north pole. The climate was becoming positively balmy. It wasn't just the warmth and increased rainfall that plants were enjoying – there was an important change in the atmosphere as well. As the Ice Age drew to an end – between 15,000 and 12,000 years ago – the level of carbon dioxide in the atmosphere rose from 180 to 270 parts per million. Experiments have shown that this would have resulted in up to a 50 per cent increase in productivity for many types of plants, and that even resilient grasses would have seen a 15 per cent increase. The rise in carbon dioxide at the end of the Ice Age didn't trigger the development of agriculture; there were so many other factors at play. But – and this is a big but – it may have been a necessary condition for agriculture to emerge – and perhaps explains why this development of human culture didn't happen any earlier, in the Ice Age.

As the world warmed and plant life flourished, grasses presented a dependable source of nutrition. As carbon dioxide levels rose in the atmosphere, the number of grains per plant would have increased, and stands of wild cereals would have grown in size and density – natural fields just waiting to be harvested. The choice of wild grasses as a food source starts to look less surprising – these were stable, dependable and plentiful resources. And, for a while, the earth was rich in its bounty.

And then there was a hiccup. A fairly large hiccup – in the form of just over a millennium of winter. The downturn in global climatic conditions is known as the Younger Dryas. This rather obscure-sounding name refers to a flower: the eight-petal mountain-avens or *Dryas octopetala*. This pretty, evergreen dwarf shrub with simple, white rose-like flowers loves the cold. If you have layered lake sediments, going back through thousands of years, and some layers contain lots of *Dryas octopetala* leaves, then you know that layer formed when the land around was alpine-tundra. Scandinavian lake-beds have deeper layers with mountain-avens leaves, dating to an older, shorter cold snap around 14,000 years ago – the Older Dryas. Then there's a later, thicker layer, dating to between 12,900 and 11,700 years ago – the Younger Dryas.

In the Middle East, this global cold snap manifested itself in reduced rainfall – and cold enough winters for frost to form. Food resources must have been severely affected. So perhaps it was with an air of desperation, in this period of relative drought and cold, that people tried to control their food supply – to grow the crops they had begun to depend on, rather than merely to gather them.

While the cooling of the Younger Dryas might have pushed people towards cultivating crops, it's possible that the warmth and bounty of the preceding millennia may have contributed to a change which made the deprivation of the cold snap even more acute. When the world had begun to warm up after the last glacial maximum, the human population began to boom. This is *before* agriculture emerges. It's possible that the expanding human population may have somehow driven the change from hunting and gathering to farming – rather than the other way around. Perhaps the booming population was already placing resources under some pressure, just as the Younger Dryas loomed.

The post-glacial baby boom wasn't the only change amongst populations of *Homo sapiens* in the Near East – society itself was changing. The most striking evidence of this can be seen at a breathtaking archaeological site in southern Turkey, in Upper Mesopotamia – a site I was lucky enough to visit in 2008: Göbekli Tepe. I described it, back then, as 'the most spectacular archaeological site I have ever seen' – and it still is. I was given a tour by the director of excavations there, the German archaeologist Klaus Schmidt, who passed away in 2014, aged sixty. And so my memory of visiting Göbekli Tepe, with Klaus as my generous-hearted guide, is now tinged with sadness. He was so dedicated to this place – and the tale it had to tell – and keen to share its story with others.

He made his discovery in 1994, while surveying the landscape for potential Palaeolithic sites. 'I was suspicious when I first saw this site: no force of nature could make such a mound of earth in this location,' Klaus told me. And he had been right to be suspicious: the mound was a 'tell', created by the accumulation of Stone Age ruins, rising some fifteen metres above the limestone plateau on which it sat. When Klaus started to investigate, he uncovered large, rectangular blocks of stone that couldn't be moved. When he dug down deeper, he found that these blocks were just the tops of giant T-shaped standing stones, arranged in a circle. When I visited, Klaus had excavated four of these circles, but he believed there were many more, still buried in the rubble of the hill.

I was blown away by the sight of these stone circles, as Klaus led me to the top of the hill and we looked down on one, in the trench below us. The standing stones were indeed huge, but they were also decorated. There were low-relief carvings – of foxes, boar, leopard-like creatures, birds, scorpions and spiders – on the sides of some of the stones. But there were also 3D sculptures, carved in one piece with the standing stones – one of a wolf, crouching on the short side of a pillar, and another of a ferocious, fanged animal head. Some stones were carved with more abstract forms, with geometric, repeating patterns. Klaus pondered the meaning of these carvings – could the animal forms represent different clans, or elements of a lost mythology? Or perhaps the guardians of the megalithic circles? He saw the images as pre-hieroglyphic communication – they clearly held meaning for the people who had made them, even if that meaning is now lost.

Although Göbekli Tepe is unique, there are echoes of the architecture and images at other sites. Similar T-shaped pillars were discovered at the ancient settlement of Nevalı Çori, and at three other sites nearby. Similar iconography – including depictions of snakes, scorpions and birds – is seen on shaft-straighteners from Jerf el Ahmar and Tell Qaramel, and on stone bowls from Çayönü, Nevalı Çori and Jerf el Ahmar. Across this part of Mesopotamia, these people were clearly connected by shared, complex rituals and mythologies.

A few stones were carved with huge, arm-like limbs that ended in clasped hands, with interdigitated fingers, on the front edge of the pillar. There were no other human features on these pillars, just the arms and hands. 'Who are these beings made of stone?' Klaus asked me, rhetorically. 'They are the first deities depicted in history,' he said – and he was probably right.

Up on the hill at Göbekli Tepe, based on geophysical surveys which provide clues beyond the archaeologists' trenches, there were probably twenty of these megalithic, monumental stone circles. But there were no signs of habitation, such as hearths, there. This seemed to have been a place where people gathered, to build monuments, to feast and worship – but it was not where those people lived.

What makes Göbekli Tepe so gob-smackingly remarkable is its date. It was built 12,000 years ago. By hunter-gatherers, not farmers. And it's certainly put the cat among the pigeons for theories about the development of human societies at the dawn of the Neolithic. The traditional story went something like this:

An expanding human population requires more food;
people adopt agriculture to fulfil this need;
agriculture facilitates the accumulation of food surpluses;
food surpluses are controlled by just a few powerful people – complex, stratified societies are born;
these new power structures are underpinned by a new invention: organised religion.

Göbekli Tepe is clearly a monumental problem for this sequence. In this corner of Upper Mesopotamia, at least, a complex society emerged in a hunter-gatherer context. Klaus believed that Göbekli Tepe provided unprecedented evidence of division of labour. 'It's very clear we must change our ideas,' he told me. 'Hunter-gatherers don't usually work in the way we understand work.' But things were clearly different at Göbekli Tepe. 'They started to work in quarries. They started to have engineers to work out how to transport and erect the stones. There were specialists in stone-working, whose job was to produce sculptures and pillars from stone.' For Klaus, Göbekli Tepe was concrete evidence of a society that had powerful, visionary leaders; that could assemble a workforce; and that could support artists. And it's very hard to interpret the massive, decorated stone circles as anything other than a manifestation of organised religion. A fully fledged cult indeed – with powerful symbols, rich with myth and meaning for the temple builders. Before Göbekli Tepe, the idea that organised religion could have predated farming was just about unimaginable. On this hilltop, preconceptions and prejudices slid, crashing to the floor.

Even Klaus found it hard to categorise Göbekli Tepe. It was pre-Neolithic – but it was clearly something different to even the last phase of the Palaeolithic. And even something distinct to the Epipalaeolithic. Klaus was tempted to call it 'Mesolithic' – yet it was unlike the Mesolithic of northern Europe, where the term applies to slightly more sedentary, but still nomadic, hunter-gatherers. Could it be classed as early Neolithic? The traditional idea of the Neolithic package – sedentary society, pottery, agriculture – is already fractured in the Near East, with the label of 'Pre-Pottery Neolithic' used for sites where sedentism and domestication of plants and animals is demonstrated, but pottery is still to come. What could we call Göbekli Tepe – 'Pre-agriculture, Pre-Pottery Neolithic'? And if so, why 'Neolithic' at all? Faced with transitions and surprises like this, our normal categories – and any idea of packages of characteristics – break down, somewhat deliciously. History – and even prehistory – refuses to be pigeonholed as neatly as we'd perhaps like it to be.

The creation of the monumental architecture at Göbekli Tepe must surely have involved a communal effort which extended beyond a few local settlements. And perhaps there's a connection between that cooperation and another feature seen in the archaeological record at this time: evidence for feasting on a large scale. The settlement site of Hallan Çemi, occupied in the tenth century BCE, seems to be very much set up for partying, with dwellings arranged around a central courtyard littered with the remains of fires and animal bones. Göbekli Tepe itself is full of a huge amount of smashed-up animal bones – from gazelles and aurochs to wild ass. It looks as though people were congregating and feasting here again and again. Plant remains are few and far between at the site, but traces of wild einkorn, wheat and barley have been found. Perhaps the feasts involved gruel or bread as well as meat. It's even been suggested that the eventual domestication of cereals in this area could have grown from a culture which invested heavily, not in bread-making, but in beer-brewing – and that alcohol could have flowed freely, greasing the wheels of social intercourse, at these ancient feasts. Much later, the workers who built the Egyptian pyramids were paid in beer. Could there have been a similar reward for labouring at Göbekli Tepe?

The importance of feasting in the Bronze Age and Iron Age – as a form of social glue, and a way in which elite individuals could demonstrate and enhance their high status – is widely accepted. But perhaps feasting has much more ancient roots – stretching right back to the dawn of the Neolithic. The improving climate after the end of the Ice Age may have provided the opportunity for individuals to accumulate wealth – in the form of surplus food – and influence – by providing lavish feasts. The scene was set for the emergence of stratified society. And so, Klaus Schmidt and his colleagues have argued, feasting – with or without beer – could have been the key stimulus for the development of agriculture.

All of these factors are so inextricably intertwined, it's impossible to pull out one and point to it as the single reason that people began to grow fields of wheat right across the Fertile Crescent, and beyond, around ten thousand years ago. It seems that agriculture probably wasn't even a possibility until carbon dioxide levels in the atmosphere rose at the very tail-end of the Ice Age, and plants became more productive. And then expanding human populations may have placed resources under pressure, especially during the climatic downturn of the Younger Dryas. But there were clearly also changes within society as the human population expanded – which we now know preceded the adoption of agriculture. The dawn of the Neolithic in the Fertile Crescent seems to be intimately linked to the emergence of complex societies, of powerful people and powerful cults, and, perhaps, a predilection for feasting.

From the Levant to the Solent

The complexity of human society that already existed, before agriculture, before civilisation as we know it, helps us to understand how ideas – and materials – travelled around, and rippled out.

Archaeology provides us with wonderful insights into the connectedness of ancient societies. The shared iconography between Göbekli Tepe and other archaeological sites as far east as Çayönü in south-eastern Anatolia and Tell Qaramel in north-west Syria shows just how far cultural links stretched across the Near Eastern landscape: Çayönü and Tell Qaramel are about 200 miles apart. The multiple centres of origin for domesticated species across the lands at the eastern end of the Mediterranean blow the idea of a small 'core area' out of the water, but also attest to cultural connections and systems of exchange that allowed ideas – and seed corn – to travel widely. The Neolithic didn't emerge in a corner of south-east Turkey; it was born from multiple, connected centres stretching across the Middle East and beyond: domesticated einkorn has been found on Cyprus, dating to 8,500 years ago – just as early as those sites in the old 'core area' of northern Mesopotamia.

And then, just 500 years later, we have that DNA fingerprint of einkorn from the submerged Mesolithic site under the Solent. How did that wheat travel all the way from the eastern Mediterranean to the fringe of north-western Europe, all those thousands of years ago? We're familiar with the trade networks that extended across the Roman Empire, two thousand years ago. Archaeology has revealed extensive trade existed earlier – in the Iron Age, even going back into the Bronze Age and the preceding Neolithic. But long-distance trade amongst small, isolated bands of hunter-gatherers eking out a living in the Mesolithic or Epipalaeolithic? That surely seems like a step too far.

Well, it might do until you take a look at examples from much more recent history. The Native Americans of the Northwest Coast maintained trade links across a huge area, spanning hundreds of miles – exchanging goods, gifts and marriage partners. These links were fundamental to power and prestige. In Australia, pre-European colonisation, the exchange networks of aboriginal communities stretched from coast to coast, right across the continent. And archaeologists are finding more and more evidence of raw materials and finished objects being carried over long distances across Europe in the Mesolithic. Flint from the coast travels some 50 kilometres inland in

Brittany; axes made from Norwegian dolorite turn up in Sweden; flint blades from Lithuania turn up nearly 600 kilometres away, in Finland; amber from the eastern Baltic also makes its way to Finland; graves in the late Mesolithic Vedbæk cemetery in Denmark contain pendants made from the teeth of elk and aurochs – both species that were locally extinct at the time. Of course, an object may have changed hands several times over those distances. But the distribution of such widely travelled goods implies that people were moving around both by land and by sea. Archaeologists believe that Mesolithic people were making voyages of up to 100 kilometres – probably using dugout canoes with outriggers. The acquisition of exotic materials seems to be linked to the development of a change in society in northern Europe, as egalitarian hunter-gatherer communities become more interested in status. Society becomes stratified – the world's oldest class system begins to emerge. It's not quite Downton Abbey, but archaeology starts to reveal a distinction between high- and low-status individuals: rich and poor. Some elaborate Mesolithic burials from around the Baltic Sea contain exotic items – presumably acting as symbols of social status.

Just as social stratification may have played into the origin of agriculture in the Middle East, it may have eased the transition in the north and west. If you're only focused on basic subsistence, you can be quite insular. If you're interested in acquiring exotic goods, and status, then you need to connect with the wider world. And the Mesolithic people across Europe appear to have been much more connected than we've previously assumed.

Mesolithic networks of exchange – of materials, ideas and people – meant that the hunters and foragers of the west were already communicating with the first farmers in the east. By 6,500 years ago, farming groups were well settled in the Danube Valley. Hunter-gatherers to the north – still living a Mesolithic way of life – borrowed pottery, T-shaped antler axes, bone rings and combs from their Neolithic neighbours to the south. They probably traded furs and amber in exchange. Nevertheless, 8,000 years ago is still extremely early for a trace of einkorn in a Mesolithic settlement on the north-west fringe of Europe. The Ice Age had only lately loosened its grip on these northerly latitudes.

The warming climate at the end of the Ice Age had an impact on the environments of the Middle East – but an even more profound effect in north-western Europe. This was where the ice sheets had descended, holding the northern part of the continent in a frozen grip for thousands of years. To the south of the ice itself, a large swathe of land had been treeless tundra. Local populations of warm-adapted species – including humans, bears and oak trees – went extinct where the ice sheets and tundra dominated.

Cousins in the south clung on, in still-habitable refugia in the south of France, Iberian Peninsula and Italy. When warmth returned, and the ice sheets retreated, much of northern Europe was left covered with sandy sediments from the rivers that flowed from the melting ice, and finer till left by the glaciers themselves. Sedges, grasses, dwarf birch and willow colonised the fresh landscape, transforming it into steppe tundra. After the chill of the Younger Dryas had passed, by 11,600 years ago, birch, hazel and pine began to spread northwards again.

By 8,000 years ago, the landscapes of northern Europe – including the peninsula that was Britain – had become enfolded in woodland, with lime, elm, beech and oak. The forests were full of life – aurochs and elk, wild pig, roe and red deer, pine marten, otters, squirrels and wolves, and plenty of wildfowl. Coastal waters were teeming with molluscs, fish, seals, porpoises and whales. Mesolithic people made good use of these resources – they were hunter-fisher-foragers. Armed with bows and arrows, and accompanied by dogs, they hunted animals on land. Kitted out with canoes, nets, hooks and lines, and fish traps, they pulled fish out of seas and rivers.

Humans started to recolonise northern Europe at the tail-end of the Younger Dryas, reaching Britain by 9600 BCE. The first colonisers didn't even need to get their feet wet. Sea levels during the Ice Age had been as much as 120 metres lower than today. As the ice melted, the sea level rose – but the first plants and animals returned to repopulate Britain when this soon-to-be island was still firmly connected to mainland Europe.

The classic picture, emerging from the archaeology, has been of small, mobile groups of hunter-gatherers – moving frequently,

leaving little trace. Mesolithic sites are, typically, very modest in size
and only briefly occupied. But at Star Carr in Yorkshire, recent
excavations have revealed the existence of a surprisingly large
Mesolithic settlement. A platform of worked timber extends along
some 30 metres of the lake edge at this 9,000-year-old site. The
whole area covers almost 2 hectares – 20,000 square metres. In such
a large, sedentary community, it's very likely that society was
hierarchical to some extent.

Even if there were small, mobile bands of hunter-gatherers roaming
north-west Europe during the Mesolithic, it seems that – at least in
some places – human society was becoming more complex. In this
context – with the existence of larger, more settled, more complex
and more well-connected groups than has been previously assumed –
the discoveries at Bouldnor Cliff seem less astonishing perhaps.

Lying at opposite ends of England, Star Carr and Bouldnor Cliff
both suggest that we've probably underestimated the complexity of
the early Mesolithic in Britain. And, just as in the Middle East,
social complexity appears to have preceded the adoption of
agriculture, not to have flowed from it. Mesolithic lifestyles were
quite variable: some communities seem to have been quite sedentary;
others were developing seafaring – shown by the trade in obsidian
around the Mediterranean as well as evidence for deep-sea fishing.

And yet the 8,000-year-old einkorn DNA at Bouldnor Cliff still
seemed like a bolt out of the blue. Traditional methods of investigation
– archaeology and botany – show domesticated einkorn appearing all
over Mesopotamia, and spilling over into Cyprus, nine to ten thousand
years ago. The Neolithic rippled across Europe, from east to west,
reaching all the way to Ireland by some six thousand years ago.
Einkorn was being grown in the middle Danube basin by 7,500 years
ago. Einkorn had reached Switzerland and Germany over five thousand
years ago. But the spread of the Neolithic seems to have been even
more rapid along the Mediterranean coasts. Particular types of pottery
seem to arrive as part of the Neolithic package at early sites in western
Europe, spreading along the coasts. Recent excavations have revealed
the presence of Neolithic farmers as far west as the south coast of
France – by 7,600 years ago. These early French farmers had pottery,

domestic sheep, emmer wheat – and einkorn. With definitive evidence of einkorn in France just 400 years earlier than the trace at Bouldnor Cliff, the gap seems to be closing. No one's suggesting that the Bouldnorites were early farmers – but they were connected to the wider world. Agricultural produce from the near continent was making its way into Britain before the arrival of farming itself.

The tale of the einkorn at the bottom of the sea opens our eyes to new possibilities, and surely reminds us not to be dogmatic about our reconstruction of the past. It's difficult – if not impossible – to find the earliest example of anything, anywhere. Genetics is now joining the armoury of archaeological tools and enabling us to draw out the tiniest, buried clues. Dates are being pushed back. That taste of wheat, perhaps even of bread – of a new way of life – had reached the coast of southern England earlier than anyone had thought possible.

Imagine being a Mesolithic hunter-gatherer, camped at Bouldnor. One day, you're visited by voyagers, people from a faraway tribe who you see from time to time. When they arrive, you offer them hospitality – they sit down to eat a meal of roast reindeer meat with you. They bring something else to the table – something quite different to the foods you could gather near home: hard little seeds. The visitors show you the seeds can be ground down, mixed with water, rolled and flattened in the hands, then cooked on the flat stones in the hearth. You eat something new and delicious that night: flatbread. This is what people at the other end of the long sea eat all the time, the voyagers tell you. These little seeds came originally from vast grasslands in the Sumerlands, the land where the sun rises.

We'll probably never know how that wheat got to Bouldnor, and indeed whether it was made into porridge or bread and eaten there. But it certainly makes you wonder whether those Mesolithic hunter-gatherers knew anything about this other way of life, that was inexorably creeping closer, along the coastlines of Europe. Could they even have imagined that the grains in that bread had been deliberately grown rather than gathered? And yet there would come a time, not too many centuries hence, when even the forests of Britain would give way to fields.

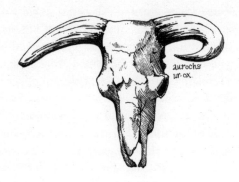

aurochs
ur-ox.

CATTLE

Bos taurus

Brindled cow, bold freckled,
Spotted cow, white speckled;
Ye four field sward mottled.
The old white-faced,
And the grey Geingen
With the white bull
From the court of the King,
And thou little black calf, suspended on the hook,
Come thou also, whole again, home.

Twelfth-century Welsh poem

The riddle of the long-horned beast

I write anywhere I can. I take my laptop everywhere with me, and I write on trains, planes and in taxis. I write in hotel rooms when I'm off for meetings or filming. I sit in cafes and write, when I'm away in cities. But the place where I feel that writing flows out of me best is at home. I sit in the bay window of the cottage and type away. I can glance up at my garden, filled now with splashes of early-autumn colour – with all manner of domesticated species that I have planted there for no other reason than that they look beautiful. The echinacea and rudbeckia are in flower – like gems of yellow and purple-pink amidst the greenness. My roses are blossoming again, clambering over the rose arch and clinging on to a lingering warmth.

The field beyond my garden stretches off, framed in the distance by copper beech trees – more dark purple than coppery now. And in that green field, dark shapes moving in the morning mist, are the cattle. They hang together in a loose herd, eating and eating their way through the day, tearing up the lush green grass that has grown again after haymaking. They're all young males. Sometimes startled and rushing from one end of the field to the other. But mostly quite still and serene. I look up as I'm trying to straighten my thoughts out and pull together the threads of this story, and I find their presence calming.

Even though these cattle are actually young bullocks, a few with quite formidable horns, I'll walk across their field without too much trepidation. They rarely show much interest in humans in their environs – unless it's the farmer in his truck. Later in the year, as autumn tips into winter, he'll drive his Hilux into the field and throw hay bales off it. The bullocks will run over to follow the truck, eager for the sweet taste of old hay. They can move fast when they want to. But mostly they are still, or moving slowly, mowing the grass, step by step. I won't take a path through the herd – that would be foolish. But I'm happy to venture into their field. Only on one or two occasions have I felt intimidated enough by a specific bullock to retreat, slowly, back to the gate and out.

These creatures are huge compared to me – ten times larger and heavier, at some 600 kilograms. Adult bulls can weigh twice that. But the ancient ancestors of cattle – the aurochsen, 'ur-oxen' – were even bigger, with the largest estimated to have weighed as much as 1,500 kilograms. You have to respect the audacity of any hunter-gatherer prepared to take that on: not just to hunt these huge animals, but to catch them and try to tame them. The skull of an aurochs on display at the Museum of London, with its formidable, metre-wide pair of horns, makes that moment of sheer, crazy bravery even more astonishing.

Our Ice Age ancestors shared the landscape with these huge beasts. Hunting them is one thing – but how did such an enormous, intimidating animal ever become domesticated?

The Formby footprints

I drove over the dunes in my old VW van – a Type 25 Syncro equipped with four-wheel drive – and down on to the beach at Formby. This was my trusty camper van, bought from my dear friend and mentor, archaeologist Mick Aston. It was pretty on the inside – I'd painted Hokusai-style waves on the plywood interior. The outside was good-looking too, in bright metallic green. But it meant business: it had a sump protector; it had that four-wheel drive that could get you out of large holes on beaches (tried and tested); its cousins had crossed the Sahara.

And so – with the blessing of the National Trust – I had no qualms about driving my van over the dunes and down on to the beach. I followed the ranger in his Land Rover. The van grunted a little as we drove up the side of a dune; I could feel the power shifting, but the wheels didn't spin. We were filming there – the very first series of *Coast* for BBC2. Production had not reckoned on the wind and rain that might blight our filming. The van became a refuge. It was warm and dry. I could even offer the crew and contributors cups of tea, brewed up on the tiny gas hob.

When the day brightened up, we emerged out of the van to survey the beach and plan our capture of it for the small screen. It's such a vast stretch of sand. It was almost impossible to see to the ends of it.

Formby Beach is the southern continuation of Southport Beach – where, on 16 March 1926, the ex-fighter pilot Sir Henry Segrave broke the world land speed record, in his bright red, four-litre Sunbeam Tiger, nicknamed Ladybird. His top speed was just over 152 miles per hour. Just a month later, the record was broken, but Segrave won it back in 1927, smashing it again in 1929 – this time at Daytona Beach in Florida. The photographs from the race day in 1926 capture the excitement of the moment, with a crowd gathered on the beach. Some onlookers climbed up on the dunes to get a better view of Segrave in his Ladybird.

But it's not just racing cars that tear up the sand here. Every spring tide brings with it energetic waves which surge right up and crash on the sand, clawing it back, exposing deeper layers of sediment underneath. It was these deep layers that had drawn me to Formby. I'm peripherally interested in raw geology, but when the traces of animals and humans start to appear in these media of sand and mud, silt and stone – that really piques my interest. And this is why I was there – to film those ancient silty sediments, under all that fine sand. A land speed record from ninety years ago was too recent to register. A blink of the eye; just yesterday. What I wanted to see was traces of the past from thousands of years ago. I knew they were here, on this beach.

In March 1989, retired schoolteacher Gordon Roberts was walking his dog along the sandy shore when he noticed strange impressions in newly exposed, deep silty layers. They were the right size and shape and spacing to be footprints. He looked more closely. That's exactly what they were. Gordon found more and more footprints. He wasn't completely taken by surprise, as local people knew that these traces turned up from time to time on the beach. But, perhaps rather curiously, no one seemed to have taken much notice of them.

Gordon brought the footprints to the attention of archaeologists, who were able to determine when they'd been formed, using various techniques including radiocarbon dating of organic fragments in the silt. They dated to between 7,000 and 5,000 years ago – an interesting period in our prehistory, spanning that crucial transition from the Mesolithic to the Neolithic in Britain.

The footprints – once exposed by a high tide – would be quickly washed away, remaining only a few weeks at most. Having realised their antiquity and their importance, Gordon decided that he should try to preserve this rare and precious data – and he embarked on a huge personal project: to record the footprints. He drew and photographed them. When he came across particularly well-preserved prints, he would make plaster casts of them. His garage began to fill up with boxes and boxes of footprint casts. When I met Gordon on Formby Beach, back in 2005, he'd recorded over 184 trails of human footprints – both male and female, adult and child. He showed me some of the photographs and plaster casts he'd made. Some contained astonishing detail of toes and the pressure of the footfall. How were such finely moulded footprints created – and preserved for all those years?

The environment around Liverpool Bay, at the time when the footprints were made, would have been very different to today. There were no waves crashing on to the beach at high tide. The sea level was lower and a long sandbar existed, beyond the coast. Behind the sandbar, there was a tidal lagoon and a gently shelving, muddy beach – which would have been largely inundated at high tide, but with gently rising water, rather than high-energy, crashing waves. Pollen analysis conjures up a large area of saltmarsh behind the mudflats with sedges, grasses and reeds, fringing into fen woodland, with pine, alder, hazel and birch. There's no reason to suppose that our ancient ancestors wouldn't have enjoyed a trip to the seaside, just as much as we do, but it would also have been a rich environment for those Mesolithic hunter-gatherers to exploit. The archaeology of the area at this time shows a concentration of activity along the coast and river valleys, which drops off inland. This is a familiar picture. Many traces of Mesolithic people are found around the coast, and beside lakes and rivers – it seems that these liminal environments had more to offer hunter-gatherers than the increasingly dense forest of the interior of Britain. At Formby, that thick forest would have started about a mile and a half inland from the coast, with its saltmarsh, mudflats and tidal lagoon.

Looking at Gordon's plaster casts, I could see how squelchy the mud must have been, to preserve those details of heel, arches and toes. The adult footprints have the splayed toes characteristic of unshod people. But in order for a footprint to have any chance of surviving, the mud can't have stayed wet and impressionable – it must have baked hard, on a hot day. Then, when the tide came in again, the gently rising waters would have brought a fine layer of sand and silt, to settle over the footprint. Again and again, until it was buried deep under silty layers, sealed and preserved. In the intervening millennia, the dunes moved back and covered the layers of silt containing the footprints. Now the dunes are retreating even further, exposing those deep silts to the raw energy of the Irish Sea – and the layers are being stripped away until the footprints are revealed.

Footprints are rare in the archaeological record, and provide a unique insight into human behaviour. With the help of experts in anatomy and locomotion, Gordon mapped out the ancient visitors to the beach: women walking slowly along the shoreline – perhaps gathering razor clams and shrimps; men running – perhaps hunting; children running around in circles, mudlarking – just like children playing at the seashore today.

But as well as the human footprints, there were animal prints. These recorded the rich bird life of those mudflats – with the prints of birds such as oystercatchers and cranes specifically identifiable. And then there were mammals: boar, wolves (or large dogs), red and roe deer, horses – and the unmistakable cloven hoofprints of aurochsen: wild cattle.

As I walked along Formby Beach with Gordon, on that cold and windy day of filming, more than ten years ago now, we kept our eyes on the ground – looking for freshly uncovered prints. It wasn't long before we found traces of aurochs hooves. They were hard to miss – they were *huge*. They were also deep – you could almost feel the great weight of the aurochs pressing its feet down into the wet mud. We both crouched down to look at the print more closely. I'm used to seeing the hoofprints of cattle – the bullocks in the field at home congregate around their water troughs, which become surrounded by a sea of fluid mud when the weather is damp. Sometimes there are days when the weather conditions are perfect for preserving the prints for a short while – a shower to slicken the mud, then hot sun to bake it hard, with all those hoofprints in it. But this aurochs print was easily twice the size of those bullocks' hoofprints.

As well as being so large, the oldest of the cattle prints at Formby are firmly Mesolithic. So these are not domesticated cattle being driven down to the coast to graze, just as cattle were driven down on to the fens in Norfolk in the Middle Ages. These hoofprints are too early to be those of domesticates – they were clearly, definitely, made by the wild predecessors of our modern cattle.

It was bleak that day on the beach. But beautiful too. We continued to film as the shadows lengthened and the dunes were momentarily bathed in golden light, just before the sun sank into the sea. We finished up, packed the kit into the back of the green van, and I thanked Gordon, before driving off across the dunes.

And Gordon Roberts – he continued collecting his records of the footprints on Formby Beach – making sure those ephemeral traces were catalogued and preserved. He died in August 2016, leaving that legacy behind – a wonderful archive for researchers of the future to delve into. I feel very privileged to have met him, to have walked with him and looked for footprints, on that long stretch of silty sand and ancient mud, by the perpetually shifting dunes.

Hunting the aurochs

The appearance of human footprints alongside hoofprints of deer and aurochsen on Formby Beach has led some researchers to conjecture that people were hunting these animals in the reedbeds and mudflats along the coast. Herds of deer and aurochsen – out in the open – would surely have attracted Mesolithic hunters. It seems like an entirely reasonable suggestion, but, unfortunately, it's impossible to tell if the human and animal prints were made at near enough the same time, on a specific day. After all, I can walk into the field at home and leave my footprints in the mud, hours after the bullocks have moved off.

What's really surprising, though, is *where* the prints were found. It's not unusual to find modern red deer on the coast, but the aurochs has long been thought of as a forest animal. And yet the wild cattle at Formby weren't just browsing along the edges of the fen woodland – they were clearly venturing right out into the open reedbeds of this coastal, wetland environment. Not the shy creatures of the forest that we once believed them to be, then.

There may be no direct traces of Mesolithic hunters stalking aurochs at Formby, but there's plenty of evidence at other sites elsewhere in Britain and north-west Europe. Most of this evidence comes in the form of butchered aurochs bones, from numerous Mesolithic sites – including Star Carr in Yorkshire. Earlier, Palaeolithic sites record the same taste for wild beef. And at just a few, rare sites, there's evidence of the hunt and the kill itself.

In May 2004, an amateur archaeologist in the Netherlands came across a curious scatter of pieces of bone and two fragments of a flint blade, all sitting on the surface of the ground, close to the River Tjonger and the Balkweg road in Friesland. The artefacts had apparently been brought to the surface by the recent digging of a ditch, and the bones had been lying exposed for some time – they'd been bleached white by the sun.

This stretch of the River Tjonger has been domesticated – its wild, winding course has been constrained into a canal. But the artefacts came from a sandy sediment which had once formed the bank of an inside-bend of the river, deep in antiquity. The digging of the ditch had completely destroyed the original site of the bones and flint – they were, in archaeological parlance, out of context – yet they could still provide some useful information.

The bones were from the spine, ribs and feet of an aurochs. They seemed small for aurochsen, but the radiocarbon dates came in at around 7,500 years ago – the late Mesolithic, too early for domestic cattle. The first domestic cattle arrived in the Netherlands at least a millennium later. And the spines of the vertebrae, sticking up from the vertebral bodies like fins, were long like those of aurochsen – much longer than those of domestic cattle. The bones of the feet were also more aurochs-like – long and slender. The final interpretation was that the bones had belonged to a small, female aurochs.

So – a dead, ancient cow. Still nothing much to write home about – except that eight of the bones bore cut marks: evidence of butchery. There were also traces of burning on some of the vertebrae.

Humans had clearly interacted with this carcass. The two fragments of flint found with the bones fitted together to form a single blade – in all likelihood, one of the tools used to skin and butcher the dead aurochs. Like some of the bones, the flint blade was burned. The Mesolithic hunters had lit a fire, perhaps even cooking and eating some of the meat right there, before carrying off the rest of the carcass, including the head.

Balkweg is just one of a handful of sites which record human interaction with what must have been a whole aurochs carcass – a single animal that is presumed to have been brought down in a hunt. There are a couple of other sites in the Netherlands, two in Germany, and one in Denmark, that seem to show the same thing: the finale of a successful hunt. Plenty of individual aurochs bones and fragments of bones also turn up in places where people were living – meat taken home for dinner. But even at these sites, only a very small percentage of the total number of animal bones are ever from aurochsen. Numbers can be misleading. The aurochsen were such huge animals, the meat to be had off one aurochs thigh would have massively outweighed that on the equivalent part of a beaver, badger, boar, or even a deer. And whereas hunters may have brought a whole boar back to the camp for the family to eat, they're unlikely to have attempted to carry an entire aurochs home. The carcass would have been jointed *in situ*, dividing up the limbs into manageable portions to transport back, along with the skin of the animal. The hunting sites show that the feet – with slim pickings – were often left behind.

The Balkweg aurochs seems surprisingly small – estimated to stand just 134 centimetres tall at the withers, or shoulders. So she seems, potentially, to have been a less formidable target for those Mesolithic hunters, and she also raises the possibility that many, later aurochs have been misidentified as domesticated cattle, or as hybrids with aurochs – perhaps an easy mistake to make if osteologists look at size alone.

Nevertheless, we know – from the date – that the 7,500-year-old Balkweg cow must have been an aurochs – a member of this ancient species, *Bos primigenius*, whose huge herds ranged right across Eurasia, from the Atlantic to the Pacific coasts, and from India and Africa in the south to the Arctic tundra in the north. Hunted by humans and other predators, the aurochs would eventually go extinct. But there were still aurochsen alive in Roman times. In the sixth book of his epic *Gallic Wars*, Julius Caesar described these *uri* – wild beasts inhabiting the Hercynian Forest of southern Germany:

These are a little smaller than an elephant in size, and of the appearance, colour and shape of a bull. They possess great strength and great speed; they spare neither man nor wild beast which they have caught sight of. The Germans trap these beasts in pits and kill them. The young men harden themselves with this labour, practising this type of hunting, and those who have slain the most, producing their horns in evidence, are highly praised. But even when very young, these beasts cannot be habituated to humans or tamed.

It's a fantastic portrait – of the wild Germans of the great Forest, as well as the formidable *uri* themselves – these untameable, horned monsters.

And yet, of course, we know that some of these magnificent animals *were* tamed. Although we talk about the species having gone extinct, some lineages survived. The descendants of the aurochs that would make it through to the present day were the ones who became allies of humans. Even as the Balkweg aurochs met her fate on the banks of the ancient Tjonger, on the north-west edge of Europe, some of her cousins in the east had already become domesticated. And not just for their meat and their hides – the same resources that made the aurochs such a prize for the Mesolithic hunters – but for their *milk*. The relationship between humans and cattle was changing.

Antelope milk and the great unbrushed

We're all so used to the idea of drinking milk now that it's difficult to stand back and try to imagine what it was like coming up with the idea in the first place. But if you can manage to shed your familiarity with milk and dairy products, then the idea of drinking another mammal's milk starts to seem very odd indeed.

Possessing mammary glands, which produce milk, is a defining characteristic of mammals. Milk is produced by females in order to feed their offspring. It's a brilliant survival strategy – it means that the mother doesn't have to abandon her infants in order to find food for them. She can stay with her brood and feed them directly from her own body. When they grow older, more capable and independent, they will be able to leave her side and find their own food in the environment.

I think very few people would be comfortable with the idea of pouring human milk on to their breakfast cereal or into their tea – but it's perfectly acceptable to imbibe the milk of another mammal. And we've been doing it for millennia. But who came up with this idea of squeezing milk out of another mammal's mammary glands in order to drink it?

I suspect that the hunting and gathering forebears of the first farmers had already tasted milk. Nobody has yet found any evidence for humans ingesting milk before the Neolithic, but that may be because no one's looked, and because it would have been a rare event. Having spent time with several different, modern hunter-gatherer communities, I've had a chance to witness just how comprehensively they can approach devouring a carcass. After a successful hunt, it's not just meat that's on the menu – offal, brains and stomach contents are all tasty and nutritious. In Siberia, I watched reindeer hunters cutting into the belly of a reindeer they'd just killed, cutting out pieces of the still-warm liver and eating those, raw, while dipping a cup into the cavity to bring out the blood to drink.

Anthropologist George Silberbauer, who spent more than a decade living amongst Bushmen in the Kalahari Desert in Botswana, described in great detail how these hunter-gatherers would utilise the carcass of a hunted antelope – including the udders: 'The udders of lactating larger antelope are regarded as delicacies when baked over an open fire. If there is milk in the udder, it is squeezed out and drunk before flaying commences.'

A traditional story from the Central Plains of North America suggests that antelope udders and milk were considered to be a prized delicacy amongst hunter-gatherers there too. After hunting and killing a doe antelope, two Kiowa chiefs were said to have argued over who should have the 'milk bags'. One of the chiefs claimed both udders, and the other chief was so shamed by this affair that he upped sticks and took all his relatives with him, heading off to new territory in the north. The faction apparently became known by a name which translates as 'Antelope Milk Drags Heart on the Ground Move Off

People'. The selfishness of the remaining chief seems too trivial to provoke such a split in a tribe; the story really seems to be about loss of power and prestige – symbolised by that refusal to share the prized udders of the antelope.

With these historical and near-contemporary examples of hunter-gatherers drinking milk from hunted animals, it seems reasonable to suggest that ancient hunter-gatherers would have done the same. They would surely have utilised such carcasses in a similarly thorough way, making the most of this precious resource. It seems foolish to suggest that no one would have tasted milk before animals were domesticated. Milk would not have formed an important part of a hunter-gatherer diet, but neither is it likely to have been completely absent. New advances in archaeological sciences provide us with opportunities to explore the diets of our ancestors, and when it comes to milk, there *could* be clues lurking in our ancestors' teeth.

Calcium is essential for healthy teeth and bones, and milk is an excellent source of this element. Like many elements, calcium exists naturally in a few slightly different forms, or isotopes. The ratios of these isotopes can be measured from samples of human and animal tissues, including bones and teeth. Ratios of carbon and nitrogen isotopes have proved to be useful indicators of diet – carbon isotopes can indicate broadly which types of plants an organism ate during its lifetime, while nitrogen isotopes reflect whether the diet was more plant-based or meat-based, and whether it contained marine sources of food. And so, for a while, archaeological scientists held out hope that calcium isotope ratios might provide clues about milk and dairy products in ancient diets. They tested archaeological animal bone and human bone, and found a difference in calcium isotope ratios between humans and other animals. But, rather disappointingly, they found no change within humans, over time. Mesolithic and Neolithic humans – living without and with domesticated cattle, respectively – had the same ratios of calcium isotopes in their bones – so unfortunately it seems that this analysis isn't going to provide us with any answers here.

Teeth do provide us with another option, though. On the whole, our ancient ancestors had much better teeth than we do today. With less sugar in their diets, they didn't suffer so badly with dental caries. You see the occasional cavity in archaeological teeth, but it's nowhere near the epidemic proportions seen in contemporary, Western society. On the other hand, our ancestors were notoriously bad at brushing their teeth. This lack of attention to dental hygiene leads to a build-up of plaque, which, over time, mineralises and becomes very hard. Accretions of hard calculus are often seen on archaeological teeth, and it doesn't stop there. Calculus leads to an irritation of the gums, and also affects the underlying bone – which starts to shrink back, until, eventually, the tooth falls out. By that point, of course, the tooth is highly likely to be lost to archaeological science. It's the teeth which end up in the grave, still *in situ* in old jaws, and plastered with calculus, that are now offering up some exciting clues about ancient diets.

As calculus forms, it traps tiny particles of food within it. At the smallest level, these include starch granules – tight packages of stored sugar – and plant phytoliths – which are microscopic, silica-rich structures that help to provide support in living plants. In the lab, these particles can be analysed and identified. Calculus studies have revealed all manner of surprising details about ancient diets. Thanks to their dirty teeth, we now know that, forty-six thousand years ago, Neanderthals in what is now Iraq were eating cooked cereals, probably barley; that the Easter Islanders ate sweet potatoes; and that people in prehistoric Sudan ate a plant called purple nut sedge, regarded as a weed today.

This is all very well, but what about the presence of milk in human diets? There are no microfossils in milk – but there are highly characteristic molecules, and one of them proves to be an essential clue. It's milk whey protein, or, more formally, β-Lactoglobulin – BLG. And importantly, for archaeologists, BLG is present in animal milk, but absent from human milk. It's also relatively resistant to being destroyed by bacteria, so it tends to stick around for a long time. Another useful feature of this protein is that it varies between species – it's possible to tell the difference between BLG from cattle, buffalo, sheep, goat and horse.

In 2014, an international team of researchers published their work looking for BLG in a range of archaeological samples. They found plenty of BLG in the calculus from Bronze Age teeth – going back to 3000 BCE, from both Europe and Russia – from cattle, sheep and goats – where there's plenty of evidence for dairying, and none in Bronze Age teeth from West Africa, where there isn't. So far, so good. This BLG study also cast some light on why Medieval Norse sites in Greenland were finally abandoned. Other studies – of nitrogen isotopes, no less – have suggested that, over 500 years – during a period of deteriorating climate – the Greenlandic Vikings were shifting to eat less food from domestic animals, and more from marine sources, including seals, before finally abandoning their settlements in the fifteenth century CE. Fish bones are often poorly preserved on archaeological sites, and it's likely that the later Vikings were eating fish as well as seal. Rather than being pathologically inflexible about their diet, as the scientist and writer Jared Diamond suggested in his book *Collapse*, it seems the Greenland Vikings were trying to adapt. Whatever the reason for their abandonment of the Greenlandic colonies, it wasn't an aversion to eating food from the sea.

The analysis of the calculus on the Viking teeth reveals another dietary change. At 1000 CE, the early Greenland Vikings were eating plenty of dairy products. But four centuries later, BLG had disappeared. So they weren't eating domestic animals any longer, and they didn't even have any access to dairy products. Perhaps the collapse of their dairy herds helped to speed the end for this Viking colony. But it's also possible that the real reason for the abandonment of Greenland could have been more baldly economic. The Greenlandic Vikings traded walrus and narwhal ivory – but as supplies of African ivory started to enter the market, their goods were no longer so valuable. Time to leave this place, then, as the bottom fell out of the ivory market and where you could no longer even get a nice bit of cheese.

All of this is fascinating, and the newly unlocked potential for using β-Lactoglobulin to reconstruct ancient diets is exciting – but this latest study looked no further back in time than the Bronze Age. Very soon, I imagine someone will go looking for milk whey protein in more ancient teeth, and I'd like to think that very faint traces might pop up even before the advent of domestication and dairying in the Neolithic, amongst the unbrushed teeth of our hunter-gatherer forebears.

Potsherds and cowherds

Our ancestors didn't pay much attention to brushing their teeth, and it seems they weren't that keen on washing up, either. To date, the earliest definite evidence we have of humans drinking milk comes from the fatty residues left on the insides of ancient sherds of pots from the Near East, dating to the sixth and seventh millennia BCE. A team headed up by Richard Evershed at the University of Bristol looked at 2,225 potsherds from south-eastern Europe, Anatolia and the Levant. They found a hotspot of early milk use close to the Sea of Marmara. This study of milk and pottery drags us away from the Fertile Crescent, to the greener and lusher north-western corner of Anatolia. And it made a lot of sense: Neolithic sites in this area contain high proportions of domesticated cattle bones, and this is an area with high rainfall and lush pastures compared with most of the Middle East. The bones tell a story of their own – with plenty of young animals in the archaeological assemblages, the early farmers seem to have been rearing cattle for both meat and milk.

The results of this study of ancient potsherds seem obvious in a way, but until Evershed and his colleagues found those traces of milk fats, it was thought that dairying was a relatively late addition to the Neolithic mode of life, coming along several millennia after the original domestication of animals, and perhaps two millennia after pottery was invented. The new evidence pushes dairying right back – to the same time as the appearance of the very earliest ceramic vessels in western Asia, in the seventh millennium BCE. Is this more than a coincidence? Perhaps needing something in which to store and process milk could even have prompted the invention of pottery.

Nonetheless the earliest evidence for both milk – and pottery – still comes along some two millennia later than the earliest appearance of domestic animals, including cattle, sheep and goats, in the ninth millennium BCE. And with these techniques, clever as they are, it's impossible to know if milk was used any earlier – because you're then in a world before pottery and there are simply no potsherds for milk lipids to stick to.

Another frustration with the evidence from milk lipids on pottery is that, unlike milk whey protein in calculus, this time we don't know *which* of the possible animals the milk came from – it could have been from sheep, goats or cattle. It may be possible, however, to get to the bottom of that by carefully studying animal bones from Neolithic sites, and an investigation carried out across eleven archaeological sites in the Central Balkans did just that. Analysis of cattle bones from those sites revealed an increasing proportion of adult animals over time. On average, adult animals made up only about 25 per cent of cattle bones in Neolithic assemblages. When numbers of young cattle are high, this suggests a focus on rearing animals for meat. At later, Bronze Age sites, from 2500 BCE onwards, 50 per cent of cattle bones come from adults. The shift towards older animals suggests that 'secondary products', such as milk (and perhaps traction as well), are becoming more important. The pattern seen in sheep bones is similar. If this pattern is reflected elsewhere, this would suggest that the first cattle and sheep were domesticated for their meat, and milking of these species came along later. But the goat bones in this Balkan study revealed something different. A higher proportion of adult animals was seen right from the beginning of the Neolithic – which starts around 6000 BCE in the Balkans – suggesting that herders in this region had always exploited domesticated goats for their milk as well as their meat. As soon as they had goats, they had goats' milk.

And yet some other recently published research prompts us to be cautious about generalising from that Balkan study. There's good evidence from other sites that cows' milk was being used right back in the early Neolithic. Once again the clues come from potsherds. This time, it's all about cheese. The first step towards making cheese involves getting particles of a specific milk protein, casein, to start sticking to each other, creating a protein net which traps fat globules inside it. This mess of coagulated protein and fat is the curds. What's left behind is a thin fluid containing some soluble proteins – the whey. There are two main ways in which you can transform milk into curds and whey: you could acidify the milk, or you could add an enzyme to it – usually rennet. Heating the milk can also speed up the process.

All of this must have been discovered by accident, by Neolithic farmers trying out new recipes perhaps, or even new storage solutions. Just imagine that you're a Neolithic farmer, off herding your animals for the day, and you want to take some milk with you. Pottery is great – but a bit heavy to carry around. Instead, you decide to use a bag made from a goat's stomach. It's not such an odd idea – bags like this are often used for water. Anyway, you fill it up with milk and off you go. Later in the day, you go to take a sip of milk and something strange has happened – it's become watery, with lumps. The rennet – the enzymes sticking around inside the goat's stomach – has transformed the milk. Rather than throw it away, you take it home and show your family. They're all quite impressed by this brand-new dairy product. But it gets even better. If you can separate the curds from the whey, you've got the beginnings of cheese. You could use a cheesecloth, or a metal sieve. Neolithic people may well have used cheesecloths, or even wicker sieves; though, unsurprisingly, neither have been discovered at any archaeological site. Cloth is not generally the sort of stuff that stands the test of time. And, in the Neolithic, metal sieves were still a long way off. But there are plenty of examples of perforated pots, which have been widely interpreted as cheese strainers. Some people have suggested other uses for these pots – ranging from lamps, to honey-straining, to beer-making. Richard Evershed's team turned their attention to fifty fragments of perforated pots from Neolithic sites in Poland, dating back to as early as 5200 BCE.

They detected lipid residues on 40 per cent of these pottery sieve sherds. And in all but one, the lipids were identifiable as milk fat. It was proof of the cheese-strainer theory – and the first definite evidence of prehistoric cheese. By processing milk, these ancient people had also done the lab scientists a favour: fresh milk residues don't last long on pottery – but the fats change when milk is processed, and persist much longer. And at these archaeological sites in Poland, 80 per cent of the animal bones are from cattle. While the milk lipids could have come from cows, goats or sheep, it seems most likely that the Neolithic farmers of Poland were indeed milking cows, and making cheese from their milk. The domesticated aurochs was here to stay.

Bones and genes

The earliest archaeological evidence of domesticated cattle themselves comes in the form of bones from a pre-pottery Neolithic site called Dja'de-el-Mughara, right on the banks of the Euphrates River. It's an extraordinary site – an ancient farming village which later became used as a graveyard in the Bronze Age. Deep down in the Neolithic layers, there are a few human burials, but also carved bone ornaments, a large circular building with wall paintings – and the butchered bones of the animals that these early farmers were keeping. There, around the Euphrates, the rolling grassy plains would have provided perfect pasture for early domestic herds during spring and winter. During the parched summer months, the villagers could have driven their animals down to the river's edge, or even out on to islands, just as they still do today. From the tricky task of managing wild herds – just think of those horns – to capturing a few aurochsen and breeding from them, the farmers had started the process of domestication. Compared with aurochs, the bones of domesticated cattle are smaller, and there's less difference between males and females. There's also a difference in horn shape, which is reflected in the bony horn core that projects from the skull. This early, skeletal evidence of cattle dates back to between 10,800 and 10,300 years ago, around the same time that the first firm evidence of cereal domestication appears in the Levant. Sheep and goats, though, are thought to have been domesticated a little earlier – perhaps just a few centuries before. It seems to make sense that domestication of these animals began to happen before the domestication of crops really took off. Pastoralism – tending herds of animals – is almost a halfway house between a nomadic, hunter-gatherer lifestyle, and a settled, agricultural way of life. But the transition from hunting and gathering to pastoralism could be very swift. One site in Turkey, Aşikli Höyük, shows a change from people subsisting on a diet including a wide range of wild animals to one where sheep made up 90 per cent of the animals eaten, over just a few centuries. Whatever prompted the pre-pottery Neolithic people at Aşikli Höyük to manage those flocks of sheep, they effectively ended up with a way of storing meat – creating a walking larder – that made their food sources more reliable.

Early genetic studies suggested that sheep and goats were domesticated many times, in separate places, but all broadly within south-west Asia. In fact, it's more likely that there was a single centre of domestication for each species – but then plenty of interbreeding with wild cousins. Domestic goats come from the wild goat, *Capra aegagrus*, while sheep are the domesticated descendants of the wild sheep or Asiatic mouflon, *Ovis orientalis*. The European mouflon, on the other hand, appears to be a domestic breed turned feral, rather than an ancestor.

It looks like a similar story for cattle. For a long time, it was believed that the two main subspecies of domestic cattle – taurine and indicine – came from separate origins. Darwin certainly thought that may have been the case, writing in the *Origin*, 'I should think ... that [the humped Indian cattle] ... had descended from a different aboriginal stock from our European cattle.' And to be fair, *Bos taurus indicis*, also known as zebu cattle, do look quite distinct from *Bos taurus taurus*, or taurine cattle. Zebu cattle have a large hump above their shoulders, and a long dewlap hanging down between the front legs. They're also much better suited to hot, dry conditions than taurine cattle. Studies of mitochondrial DNA and Y chromosomes supported this idea of a separate origin for each subspecies. But a single origin makes much more sense: it seems most likely that domesticated cattle arose in the Near East, between 10,000 and 11,000 years ago, and then spread, meeting wild relatives along the way. Reaching South Asia some nine millennia ago, a significant level of interbreeding with local aurochsen could have introduced zebu genes and characteristics to domestic cattle.

The diaspora of cattle got under way very quickly. Farmers and their cattle were travelling west too; by 10,000 years ago, someone had been brave enough to put them in a boat and take them to Cyprus. By 8,500 years ago, domestic cattle had reached Italy, and by 7,000 years ago, they had spread, with the early farmers, into western, central and northern Europe, as well as Africa. Cattle had reached north-east Asia by 5,000 years ago. As sheep and goats spread out from the Middle East, they were moving into uncharted territory for caprines – there were simply no wild relatives to interbreed with. But it was different for the domesticated bovines: wild oxen ranged right across Europe and Asia, and cattle seem to have interbred with them everywhere. The first clue came from mitochondrial DNA, where unusual variants in Neolithic cattle from Slovakia, Bronze Age cattle from Spain, and in a few modern cattle as well, were all traced back to European aurochsen. More recent genome-wide analyses have revealed widespread interbreeding between domestic cattle and local, wild oxen, right across Europe. British and Irish cattle breeds, in particular, have a lot of aurochs DNA in their genomes. But we can surely only speculate on how deliberate – from a human perspective – any interbreeding may have been.

I've spent some time living with indigenous reindeer herders in Siberia, where the domestic reindeer herds are large and impossible to guard or corral. The wild herds are even larger, and – like the domestic herds – often on the move. The reindeer herders I spoke to worried less about wild animals joining their herds, than about losing their reindeer to the untamed hordes. They were always nervous when they knew there was a wild herd nearby. Their experience has made me think differently about those early farmers and their herds.

Just how carefully did Neolithic farmers tend their cattle? Did they fence them in or let them roam more freely? Did they catch and add carefully selected wild aurochs to their herds, or does the genetic introgression simply record unavoidable contact between domestic and wild animals? If this is the case – and I have no idea if it is – then this simply means that aurochs cows were more likely to join up with domestic herds than wild bulls were.

From a biological perspective, it's not surprising that domestic cattle continued to interbreed with wild populations. The two modern subspecies of cattle have often interbred to produce hybrids. In Africa, cattle DNA reveals a history of male zebu cattle being bred into herds of taurine cattle, to produce Sanga cattle. In China, taurine cattle spread into the north and indicine cattle into the south. This north– south divide is still evident in Chinese cattle today, with taurine–indicine hybrids in the middle. Cattle can also produce hybrids with other species. One Chinese cattle breed has been found to contain yak DNA – and, conversely, domestic yaks contain DNA from cattle. In

Indonesia, zebu cattle often interbreed with the local wild cattle species, known as banteng, or *Bos javanicus*.

The riddle of the shrinking cow

When they entered into an alliance with humans, cattle, sheep, goats and pigs changed. In contrast to grains of wheat, which grew larger under domestication, cattle and other animals got smaller. But curiously, cattle – unlike sheep, goats and pigs – then continued shrinking, through the Neolithic, Bronze Age and into the Iron Age. And it was a significant reduction. Archaeologists have been able to quantify the shrinking that took place during just the Neolithic by scrutinising ancient bones from European cattle, where farming got started around 7,500 years ago (5,500 BCE). By the end of the Neolithic, 3,000 years later, cattle were, on average, a third smaller than they'd been at the beginning of farming.

It's easy to leap to the conclusion that early farmers might have been deliberately selecting smaller, easier-to-handle, animals to breed from. While that may have been the case at the dawn of domestication, it seems unlikely that farmers would have continued to select smaller and smaller animals over the generations and millennia. So why were cattle continuing to shrink?

The results from the careful osteological analysis – of bones from seventy sites across central Europe – allowed the archaeologists to test different ideas about what might have caused this reduction in body size. One possible explanation could be that domestic cattle were chronically underfed – but there are no signs of cattle being malnourished. A decrease in average size could come about as a side effect of a reduction in the degree of difference in size between male and female animals. Yet while there was a reduction in sexual dimorphism right at the start of the Neolithic, that trend didn't continue as the cattle themselves got smaller. Cattle arrived in Europe some 3,000 years after that initial domestication; and, over subsequent millennia, the bones of European cattle showed a fairly consistent level of difference between males and females – and yet they continued to shrink.

Climate change can also have an effect on the size of animals. Could this be the answer? Probably not, as you'd expect the wild cattle to be affected in the same way as the domestic ones – and they aren't. Another possibility is that the apparent change in average size of cattle just reflects a changing ratio of female to male cattle. A larger proportion of adult females in a herd would fit with an increasing focus on milk production; in dairy herds, young males are often culled. This seems like a good hypothesis, then – but again it doesn't fit the evidence. The bones of the Neolithic cattle didn't show an increasing proportion of female animals. The scientists were doing a good job at rejecting hypotheses. After all that rejection, there was just one hypothesis left – and one which seemed to fit the evidence from the piles of bones perfectly.

The ancient cattle bones from Neolithic central Europe revealed not only a decrease in size, but a rising number of juveniles – suggesting, this time, an increased focus on meat production. Young cattle grow quickly. By maturity, at three to four years of age, the rate of growth slows right down. You don't gain much more meat by keeping a mature animal alive. So you cull more animals before, or just as, they reach maturity – and the proportion of juvenile bones in the middens around your settlements goes up. On its own, this still doesn't explain the size reduction in cattle – because this is a phenomenon recorded in adult cattle; the juveniles are disregarded from the sample. But nonetheless, the high proportion of subadult bones is telling us something: in such a herd, many of the cows giving birth to calves will be immature themselves. These cows – able to reproduce but still with some growing to do – will tend to have calves with lower birth weights than their mature sisters in the herd. Smaller, lighter calves tend to grow up to be smaller, lighter cattle. It doesn't mean that the European Neolithic herds weren't being milked as well, but meat seems to have been a priority – and one which meant that European cattle were 33 per cent smaller at the end of the Neolithic than they'd been at the beginning. Later on, in the Bronze Age, the proportion of sites with subadults decreased – accompanied by a small increase in the size of cattle around this time. But this was a minor blip: on the whole, cattle continued to shrink right up until the Middle Ages, and it would be some time before they regained stature – and, even then, they were never quite as magnificent as their wild aurochs ancestors.

Cattle also provided services other than milk and meat for our own ancient ancestors. Julius Caesar recorded the cultural importance of the wild aurochsen to the Iron Age people of Germany, and domestic cattle would continue to play significant roles in religious rituals and ceremonial fights. The cult of the bull in ancient Crete seems to have formed the inspiration for the myth of the minotaur. Cattle may have been drafted in as formidable, worthy opponents for heroes and matadors, but their size and strength was useful in a more prosaic way as well. They were the original tractors, used to pull ploughs and wagons. In many less industrialised parts of the world, they're still used in this way. And sometimes, they're better suited to the job than a mechanical alternative. It's impossible to drive a tractor up to the high paddy fields of Longsheng in southern China – but an ox can get there easily, and pull the plough beautifully along the narrow terraces.

The breeding and use of cattle for traction may explain another strange blip against that background trend of diminishing size of European cattle. During the Roman period, European cattle get a bit bigger – as shown by analyses of bones from sites in Italy, Switzerland, Iberia and Britain. Farmers may have been deliberately breeding and trading larger cattle, but the size increase could also reflect an injection of local, wild aurochsen genes. And perhaps larger cattle were particularly sought after at this time – as essential cow-power for the expanding wheat fields of the Empire. Nevertheless, cattle remained fairly small – much smaller than today – until well after the Middle Ages.

On the hoof

Following the initial diaspora of domesticated cattle, with the first farmers, throughout Europe, Asia and Africa, bovine populations continued to shift and blend, as people moved, taking their cattle with them. As civilisations blossomed and empires grew up, flourishing breeds of cattle were transported from their homelands to new pastures.

The mitochondrial DNA of cattle in northern Italy suggests an intriguing link with Anatolia – which seems to date to much later than the original arrival of cattle in Italy. Herodotus wrote about the sufferings of people in Lydia – modern Anatolia – during an eighteen-year-long famine. Eventually, he tells us, a large contingent of Lydians left the shores of the eastern Mediterranean and voyaged to Italy. According to Herodotus, the settlers in Italy called themselves Tyrrhenians – and went on to found the Etruscan civilisation. It's a rather romantic story, with seemingly very little in the way of other historical or archaeological evidence to back it up. But perhaps the cattle of northern Italy retain a faint genetic memory of an ancient migration from the eastern Mediterranean after all. Analysis of mitochondrial DNA from ancient Etruscan human bones has also been suggested to reflect a link between northern Italy and Turkey. It's not a clear sign of a migration, though – it may just reflect how well these regions were linked by trade and mobility. But perhaps, just perhaps, Herodotus was right after all.

Trade routes are also reflected in the genetic make-up of modern cattle. Zebu DNA in cattle in Madagascar undoubtedly reflects strong trading connections with India. But some significant movements of cattle, which show up in the genes, followed human migrations around the globe. A relatively recent introgression of zebu genes into African, taurine cattle probably reflects the Arabian expansion of the seventh and eighth centuries CE.

After the Middle Ages, we start to see an increase in the size of cattle – which could either be due to selective breeding, or perhaps an indirect consequence of relative political stability and prosperity in Europe. After all, peacetime means that pitchforks can be used, not as weapons, but for their intended purpose of lifting hay.

The bovine takeover of the Americas started at the tail-end of the fifteenth century. The first cattle to be loaded on to ships at Cadiz in 1493 – part of Columbus's second expedition to the Americas – were headed for Santo Domingo, via the Canary Islands. Horses, mules, sheep, goats, pigs and dogs also made the trip. And they were soon joined by more – every fleet after that brought additional animals to add to the expanding herds and flocks.

So there were no pre-Columbian cattle in the Americas – at least, that's the traditional view. However, there's a very real possibility that cattle may have arrived in North America some 500 years earlier, with the establishment of Viking settlements in 'Vinland' – probably Newfoundland. The Norse sagas specifically describe islands off Vinland where winter was mild enough for cattle to graze outside all year round. Still, there's no evidence that these Viking colonists left any descendants – human or bovine. The colonies were abandoned, and it would be centuries before Europeans 'rediscovered' the Americas. And despite the existence of at least one Viking Age settlement, at L'Anse aux Meadows, even the link between Newfoundland and the Vinland of the sagas is still questioned by some. On the other hand, there seems no reason to doubt the well-documented voyages of the Spanish and Portuguese. The Spanish transported cattle to the Caribbean; the Portuguese took cattle with them to Brazil – and these animals were the ancestors of Latin American Criollo or Creole cattle.

During the eighteenth century, British pioneers led the way in systematic selective breeding – and specialised breeds started to emerge. Robert Bakewell bred the large, brown-and-white longhorn – principally draught animals, but also good milkers – while the Colling brothers produced the red or roan British shorthorn, good for meat and milk.

Cattle breeders engineered crosses between particular breeds to bring out desired characteristics. The nineteenth century saw a period of bovine 'anglomania' where British shorthorn bulls were bred into continental European cattle. Productive breeds from Holland, Denmark and Germany were also exported to other European countries, and to Russia, to improve domestic herds. Hardy Ayrshire cattle from Scotland were bred into Scandinavian populations. There was a mass introduction of zebu cattle into Brazil in the nineteenth century, to improve the existing herds there. Most of the milk produced in Brazil today comes from Girolando cattle – an indicine-taurine cross. In fact, the founding populations seem to have been part-zebu already – reflecting those already complex connections between the cattle of south Asia, Arabia, North Africa and Europe. And cattle have done well – extraordinarily well – in their new, New World habitats. In Brazil – home to cattle for less than 500 years – there are now more cattle than humans. Some 200 million people live in Brazil – and some 213 million cattle.

In the second half of the twentieth century, cattle breeding got even more technical, with the introduction of artificial insemination. Some cattle were carefully bred for maximum milk production – such as Holstein-Friesians, now the most populous cattle breed in the world. Others were – literally – beefed up, with traits for heavy musculature promoted through selective breeding. Some cattle were also bred to suit particular environments, from lush green grassland to virtual desert. But it wasn't all about productivity: aesthetic characteristics were under selection as well. An astonishing variety of cattle emerged – not quite as astonishing as the diversity within dogs, but prodigious nonetheless. From white to red to black and everything in between, short-haired to frankly shaggy, small and large, long-horned, short-horned and hornless – the variation in appearance of modern cattle is nothing short of impressive. Selection has changed over time – we now prefer cows which produce less fatty milk, and black-coated beef cattle are currently fashionable in the US. In the developed world, cattle are no longer needed as draught animals, so selection for strength and stamina, pulling the plough, has receded into history.

But selective breeding over the last 200 years – in cattle as in dogs – has created a paradox: there may be plenty of variation, phenotypic and genotypic, between breeds; within breeds, it's a different story. This narrowing down of variation has been carried out quite deliberately. For most of their history, domestic cattle were subject to 'soft selection' as farmers encouraged reproduction amongst animals that were more productive, or better suited to particular environments. And there was plenty of gene flow between emerging breeds. But over the last two centuries, breeders have focused on reducing the variation within breeds – until even coat colour became consistent. With the tight control over reproduction facilitated by artificial insemination in developed countries, the possibility of interbreeding between cattle breeds was practically eliminated. The result of this restriction on breeding, together with strong selection, is a species which actually consists of lots of separate, fragmented populations. Each one is at risk of all the problems inherent in inbreeding, including higher rates of genetic disease and infertility, and population-wide susceptibility to infectious disease. In the wild, fragmented populations with little genetic variation are the ones which are most at risk of extinction. And yet tightly constrained, industrial breeds may be – in the here and now – more productive than traditional breeds. For farmers, switching from a traditional breed to an industrial breed can be an economic no-brainer. But in the long term, it's not sustainable. Once a domestic breed has gone extinct, all the 'genetic resources' it contained are also lost. Geneticists are worried about the future of cattle – and our own food security – if the fragmentation of populations and inbreeding continues. They're worried about domestic sheep and goats too, but the situation for these animals differs from cattle as there are several species of each, and wild species still exist as well. Although cattle can hybridise with other, extant bovine species – which might be useful genetic resources in the future – the wild ancestor of cattle went extinct centuries ago.

Resurrecting the aurochs

As the global population of domestic cattle burgeoned, the numbers of wild aurochsen dwindled and dwindled. Once they had roamed right across Europe, into central and southern Asia, and North Africa. But by the thirteenth century CE, the territory of the wild aurochs had contracted until they only existed in central Europe. Aurochsen survived longest in Poland, where they were protected by royal decree, and even fed during the winter, to ensure the sport of kings. But even royalty couldn't save them in the end. Domestic cattle encroached on the habitat of the aurochsen. Cattle diseases and illegal hunting also played a role. But eventually, their demise was ensured by a lack of interest. In 1627, in the Jaktorów game preserve in Poland, the last recorded aurochs, a female, died.

The loss of these large grazers – especially as it happened relatively recently – is lamentable. There are very few species of 'megafauna' left in the world, and a large portion of the blame for their disappearance rests with us humans. In a more selfish way, losing these species also means that we've lost them as genetic resources. We can't inject new hybrid vigour into our cattle populations by interbreeding them with aurochsen. And there are also wider, ecological reasons to regret the absence of these animals in our modern landscapes. Without large grazers, wild places become uniformly forested. Nature becomes less diverse.

And this is why some cattle breeders are attempting to bring the aurochs back to life – at least, they're trying to create a new breed which will be as aurochs-like as it could possibly be. The breeders of the Tauros Foundation in the Netherlands have chosen several European breeds which seem to retain some 'primitive', aurochs-like characteristics – in size and shape, length of horns and grazing behaviour. By breeding these various types of modern cattle together, they are hoping to resurrect the phenotype – the appearance and possibly the behaviour too – of the aurochs. However, recent advances in molecular genetics could mean that it may be possible to do more than breed something which *looks* like an aurochs on the surface. It may be possible to produce an animal which is, through and through – genetically – an aurochs.

The first step towards doing that is to characterise an aurochs's genome – not just its mitochondrial DNA, or its Y chromosome, but its entire, nuclear genome. In 2015, a team of researchers did just that, sequencing the genome of a 6,750-year-old British aurochs. Taking a sample of bone powder from a humerus found in a cave in Derbyshire, they were able to extract the DNA and read the code. This animal lived a thousand years before the first domestic cattle reached Britain: it was pure, unadulterated aurochs. And when the geneticists compared this aurochs's genome with that of modern, domestic cattle, they found clear evidence of later interbreeding between aurochs and domestic cattle. A range of British breeds, including Highland, Dexter and Welsh Black, contained DNA from the ancient, British aurochs population. There was no evidence of interbreeding with this British aurochs in non-British breeds – which is important as it suggests the interbreeding really did happen in Britain, between the local domestic cattle and their wild cousins, rather than happening earlier, in mainland Europe. It adds to the evidence of interbreeding from those mitochondrial DNA and Y chromosome studies – so in a way, the ancient aurochs are still with us. Just how much aurochs DNA could there be, knocking around in the genomes of living cattle, from such ancient liaisons? If more aurochs genomes are sequenced, it should be possible to find more breeds which have these recent genetic additions from aurochs. This could be a better way of finding cattle to breed from, to 'recreate' an aurochs, than simply looking at characteristics. But both approaches beg the question – what's the real point of attempting this 'de-extinction'? Is it to breed a creature which looks like the extinct animal? Is it to create an animal which is, genetically, as close as possible to the original, lost species? Or is it to produce a new breed which could fulfil similar roles in an ecosystem to those performed by the extinct animals? What is most important in this endeavour – looks, genetics or behaviour? While there's a bit of me that would love the chance to see a real, living aurochs, the opportunity to reintroduce a missing keystone species into a wild ecosystem is a more worthy cause, and a more valid reason for attempting de-extinction.

The Dutch Tauros breeding programme started in 2008 with the explicit aim of creating something as close as possible to an aurochs, for release into wild reserves – to put back what has been lost: to resurrect the natural dynamics of ecosystems. They hope to have something very much like an aurochs, ready to be let loose, by 2025. It's astonishing to think that large, wild cattle could soon be roaming the rewilded wilderness in Europe. The stately, reddish-brown, long-horned ur-oxen that we know from Ice Age cave paintings could be back in the landscape very soon.

teosinte   maize

MAIZE

Zea mays mays

In chalky, barren lands bordered
by the sea, along
the rocky Chilean coast,
at times
only your radiance
reaches the empty
table of the miner.

Your light, your cornmeal, your hope
pervades America's solitudes …

Pablo Neruda, 'Ode to Maize'

Gateway to a New World

Maize, along with wheat and rice, is one of the most important crops in the world – a crucial source of food, fuel and fibres. And it's grown in an astonishing variety of different places. When you choose plants for your garden – whatever those plants are – you might look for species or varieties that are naturally suited to the habitat. The garden may have clay or crumbly, humic soil; it might be cold and damp or hot and dry. Some plants will tend to do better than others in it. Even within a garden, some plants will do better in darker, cooler spots, while others will flourish against a south-facing wall.

But maize, it seems, is not so difficult to please. It appears to be extraordinarily cosmopolitan. It's the most geographically ubiquitous grain. In the Americas, it grows in fields in the south of Chile, forty degrees south of the equator – all the way up to fifty degrees north, in Canada. It thrives in the Andes, 3,400 metres above sea level – all the way down to the lowlands and coasts of the Caribbean. The key to maize's global success surely lies in its prodigious diversity – in looks, habits and genes. But as a global crop, its history is incredibly difficult to disentangle. Although its worldwide expansion happened in just the last five hundred years, written sources are very vague, for instance, about the introduction of maize to Africa and Asia. DNA provides additional clues, but global trade and exchange have ensured that the genetic history of maize is a very tangled web. The globalisation of maize is intertwined with human history, following that ebb and flow – with voyages of discovery, trade routes stretching around the world, and the expansion and collapse of empires. But there's one thread that is easy to pick out from this mesh: a distinct moment in time that would ensure the globalised future of maize.

During the thirteenth century CE, Mongol emperor Genghis Khan and his successors carved out a huge territory for their Empire – stretching across Asia from the Pacific Ocean in the east to the Mediterranean Sea in the west. Nearly a century of aggressive expansion was followed by several decades of relative political stability: the Pax Mongolica, or 'Mongol Peace'. During this time, trade routes between east and west were actively protected, and business flourished. Then it all began to fall apart. In 1259, Genghis's grandson, Mongke, had died without a successor – and the great empire had already begun to fragment into separate khanates or kingdoms. Still, relative peace had prevailed and the Silk Road remained open for business. But by the end of the thirteenth century, the khanates of the Mongol Empire were only very loosely allied. In the early fourteenth century, wars between these separate states divided them, and one by one, they fell, to other rising powers across Asia. At the same time, the hideous spectre of the Black Death hitch-hiked along the routes that had once conveyed spices, silks and porcelain, and both Asia and Europe were plunged into turmoil.

And yet Europe still yearned for the spices of the east. These flavours of the Orient were highly sought-after precisely because they were *exotic*. Sandalwood, nutmeg, ginger, cinnamon and cloves were the flavours of power, the scents of status. The overland connection to the East was not only dangerous, it involved chains of middlemen, all wanting a mark-up. And so European merchants and explorers had been searching for some time for a viable sea route to the Orient – to India, the Spice Islands, Cathay and Cipangu (which we know as Japan). Africa was unhelpfully in the way. In 1488, the Portuguese explorer Bartolomeu Dias battled on, round the Cape of Storms – later renamed the 'Cape of Good Hope' – and the prospect of a south-east sea route seemed possible at last. But Italian explorer Christopher Columbus had another idea. A Florentine astronomer called Paolo Toscanelli had suggested that sailing west from Europe could be a quicker route to the Far East. Earlier that century, others had made attempts – they'd got as far as the Azores, only to be beaten back by westerly winds.

Columbus had worked as a sugar merchant, sailing out west from Europe to Porto Santo, in the eastern Atlantic, near Madeira. From the contacts he made on his voyages, he learnt that, while westerly winds predominated in the north, when you moved further south in the Atlantic, the winds mostly blew from the east. It was a risky thing to try – explorers usually preferred to sail into the wind, knowing they would be assured a safe return trip. But Columbus had a thirst for discovery – and social advancement. He didn't just want to find new territory, he wanted to claim it for himself: to be the governor of any islands he discovered, and to pass that position on to his heirs. Eventually, he secured financial backing from King Ferdinand and Queen Isabella of Spain, and the voyage was on.

In the third century BCE, the Greek mathematician and geographer Eratosthenes had reckoned the circumference of the globe to be 252,000 stadia. That's about 44,000 kilometres. The actual circumference is just over 40,000 kilometres – Eratosthenes was only 10 per cent out. But later geographers thought the Ancient Greeks might have profoundly overestimated the size of the earth. Toscanelli was one of them. And in 1492, a cartographer in Nuremberg – who had corresponded with Toscanelli – produced a small globe of the known world: an '*erdapfel*', or earth-apple. It's the oldest known globe in the world, and historian Felipe Fernandez-Armesto has called it 'the most surprising object' of 1492. And on it, the Americas are conspicuously absent. The implication is: if you set sail from Europe, heading west, you'll eventually reach Asia.

Setting off in 1492, Columbus chose to sail west from the Canary Islands, just off the coast of Morocco, with three ships. Not only was the wind filling their sails here, they were embarking on what they believed, from record of previous explorations, to be just about the right latitude to hit the famous port of Guangzhou in China. And so, heading into the unknown, the tiny fleet – the *Niña*, the *Pinta* and the *Santa Maria* – weighed anchors on 6 September. After a month – no landfall, and Columbus's fellow commanders were becoming impatient. The sailors were looking a little mutinous. The three ships changed course, to the south-west. In the early hours of Friday 12 October, a lookout on the *Niña* spied land. It was probably the island we now know as San Salvador, in the Bahamas.

Just imagine those Iberian explorers and sailors arriving on this island. This, to them, was the Indies: an island off the east coast of Asia. After so long at sea, they'd reached this idyllic place – the darkness of the deep sea changed to the clearest turquoise as they approached the palm-fringed beach. The island was lush and forested, full of promise. And although history is full of strings of happenstance and contingency, it *feels* as though it turns on this point, when Columbus sets foot on that beach – as his boots sink into the sand.

He met the islanders. They seemed not to be desperately suspicious of his motives, and instead were amicable and hospitable. How different history might have been had Columbus not met with such a friendly reception. To Columbus, the natives were humans, not monsters; they were naked and natural; they were morally pure, perhaps – but also easy to conquer. But this was not the Eastern civilisation he was expecting to encounter. There were none of the riches of the Orient here. There were crops, though. On 16 October 1492, Columbus wrote in his ship's log: 'It is a very green island, and very fertile and I don't doubt that all year round they plant and harvest *panizo*.'

When some of his companions came back from exploring nearby Cuba, on 6 November, Columbus recorded that they'd found a distinct type of cereal growing there: '... another grain, similar to *panizo*, that they call *mahiz*, and it tastes good when boiled and roasted.'

It's likely that these two cereals – on San Salvador and on Cuba – were in fact the same plant: maize. Plant scientists think that Columbus probably saw maize in flower on San Salvador, and thought it looked similar to *panizo* – sorghum or millet, something he was familiar with back home. So the '*panizo*' he described was in fact the same thing as the '*panizo*-like' grain that the Cubans called *mahiz* – maize.

And so, with those *mahiz* grains in his pocket, Columbus went on to explore other islands. The islanders, who travelled around by canoe, knew the local geography very well – and shared this knowledge with Columbus. But where was Japan? Where was China? He had high hopes of finding Asian civilisation on Cuba – but it wasn't there. There were no spices and silks. The inhabitants were quite poor – these were not the trading partners he was looking for.

He sailed on to the island of Hispaniola, now divided between the Dominican Republic and Haiti. There, he found both civilisation – at least a civilisation capable of producing stone-built architecture – and, perhaps more importantly, gold. Leaving a garrison on Hispaniola, he gathered up his trophies – including gold, of course, but also chilli, tobacco, pineapples and maize – and headed home. Battered by storms on the return journey, Columbus was forced to land in Lisbon – where he was interrogated by Bartolomeu Dias, before being released to sail on to Huelva. Although many doubted his story, he insisted to his patrons, Ferdinand and Isabella, that he'd fulfilled his contract: he'd found the eastern edge of Asia. In fact, he didn't know where he'd been – but he knew how to get back there.

He returned the following year, but the friendly reception he'd enjoyed in 1492 had turned sour. The garrison on Hispaniola had been massacred. Rumours of cannibalism proved to be true. And the climate was feverishly hot and humid. The indigenous people of the New World were not going to acquiesce to foreign sovereignty as easily as Columbus had imagined.

Columbus is, of course, a person who has inspired admiration and vilification in almost equal measures. He forged a connection which would see the empires of Europe rising to become global superpowers, while the Eden of the Americas was plundered and its civilisations destroyed. Setting foot on that beach, he sealed the fate of tens of millions of Native Americans and ten million Africans. The impact of that moment would ripple out through history. Until this point, Europe had been something of a backwater – but the establishment of colonies in the New World would change all that. The rise of the West had begun.

And the impact would be felt not just throughout human societies, around the world, but by the species that had become our allies – on both sides of the Atlantic. This contact between Europe and the Americas would quickly turn into a sustained connection between the Old and New Worlds. These supercontinents had been largely separate since the break-up of Pangaea, which began around 150 million years ago. During the Great Ice Age, the Pleistocene, the world went through repeated glaciations. And during the glacial periods, sea levels would fall to such an extent that the north-east tip of Asia would be joined to the north-west corner of North America, via a tract of land known as Beringia – or the 'Bering land bridge'. This bridge would allow some interchange of plants and animals between Asia and North America. It was the route by which humans first colonised the Americas, around 17,000 years ago. And yet the ancient, underlying theme of divergence and difference between the flora and fauna of the Old and New Worlds persisted – until the human-mediated transfer of plants and animals which started with Columbus bringing back his pineapples, chilli and tobacco in 1492. Plants and animals which had been contained and separate from each other made that leap across the pond, to find themselves facing new landscapes, new challenges and new opportunities on the opposite side. Cattle and coffee, sheep and sugar cane, chickens and chickpeas, wheat and rye travelled from the Old World to the New. Turkeys and tomatoes, pumpkins and potatoes, Muscovy duck and maize made the reverse journey.

The Columbian Exchange has been described by some as the most significant ecological event on the planet since the dinosaurs were wiped out. It was the beginning of globalisation: the world became not just interconnected but interdependent. But it had a wretched inception.

The fortunes of Europe (and, in due course, Asia and Africa) were transformed by the domesticated species brought back from the New World. Novel crops boosted agriculture and populations began to recover from war, famine and plague. But that was in the Old World. In the Americas, a scene of devastation ensued. Just as plants and animals had followed separate evolutionary trajectories on either side of the Atlantic, the pace and direction of technological change had been different in the Old World compared with the New. The Europeans possessed advanced technology: their military and maritime kit was vastly superior to that of the Native Americans. The immediate consequences of contact, with heart-stopping, dreadful inevitability, were tragic. Disease organisms were also part of that Columbian Exchange: the Europeans brought back syphilis from the Americas, while introducing smallpox there – with disastrous consequences. The indigenous population of the Americas plummeted after conquest. It was decimated: by the middle of the seventeenth century, 90 per cent of the indigenous population had been wiped out.

It's easy to focus on the power imbalance that existed between the Old and New Worlds in the fifteenth and sixteenth centuries. Human societies had developed in different ways in the Americas and in Europe, but it wasn't as though the Native Americans were entirely without technology – far from it. When it came to their exploitation of natural resources, they were clearly experts. It's wrong to see the pre-Columbian Americas as, on the one hand, a natural Garden of

Eden, and on the other, an innovation vacuum in need of European inspiration to realise its potential. Native American societies had a rich and diverse history of innovation, and the Americas contained completely independent centres of domestication. Many of the pre-Columbian societies of the Americas were large, urbanised – and already dependent on agriculture.

The Spanish explorers didn't pluck wild plants, out of relative obscurity, recognise their utility for the first time, and transform them into something which would greatly benefit humanity. What the Europeans found on the other side of the Atlantic were organisms which had already changed away from wildness, over thousands of years – which had already entered into a tightly bound, successful alliance with humans. What Columbus discovered was not only a new land, previously unknown to Europeans, but a wealth of useful, tamed animals and plants – ready-made domesticates.

Among those prizes was that cereal he'd spotted and written about, just four days after landing on San Salvador – the cereal that was not only a staple food but a sacred food for the Aztecs and Incas, whose civilisations would soon be swallowed up by the Spanish Empire: maize.

Maize in the Old World

Columbus returned home from his first voyage to the Bahamas with samples of seeds, bringing back even more on his subsequent trips. News of the arrival of maize spread quickly – reaching the Pope and his cardinals by 1493. Written on 13 November, a letter from an Italian historian attached to the Spanish court, Pedro Martir de Angleria, to an Italian cardinal, Ascanio Sforza, described the new grain:

> *The ear is longer than a hand, pointed in shape and as thick as an arm. The grains are beautifully laid out, and are a similar size and shape to chickpeas. They are white when unripe, becoming black when ripe; after milling, they are whiter than snow. This type of grain is called maize.*

A follow-up letter from Martir in April 1494 apparently accompanied a sample for the cardinal. And in 1517, maize appeared in a fresco painted on a wall in Rome. But although this tropical plant seems to have settled in well in Spain, it didn't take well to more temperate climates. Cold winters stunted its growth, and long hours of daylight in summer would have discouraged the plant from setting seed. So it seemed that, in central and northern Europe, maize was highly unlikely ever to become a dependable crop and a staple food, as it was in the Caribbean. And yet it started to pop up in increasing numbers of records – and not just from southern Europe. In 1542, the German herbalist Leonhart Fuchs wrote that it was 'now growing in all gardens'. By 1570, it was growing in the Italian Alps. It seems extraordinary that this tropical plant had evolved so quickly, adapting to the significant challenges of a temperate climate.

Careful reading of the great sixteenth- and seventeenth-century European herbals suggests that something else was happening. The writers of these botanical records tended to follow a fairly strict format: they'd list the names of a plant; then they'd describe the plant – its leaves, flowers and roots as well as its uses; its medicinal properties were laid out; and its geographic origin. The entries were accompanied by illustrations printed from woodcuts. Maize first appears in these herbals in the 1530s. But for some thirty years after that, its New World origin isn't mentioned. While the Spanish explorers were writing about this cereal that they'd brought back with them, many people seemed to think that maize had arrived in Europe from Asia. The first reference to maize in the herbals appears in the work of the German herbalist Jerome Bock in 1539. He referred to maize as *welschen korn* – 'strange grain', something new in Germany – and he thought it had come from India. The Medieval herbalists were so entranced with the classical world, it was almost as though they couldn't escape its stranglehold. Confronted with novel plants, they looked to the Ancient Greeks – especially Pliny and his contemporary, Dioscorides – for help. Surely they'd described everything: they must have the answer. The geographic confusion and conflation that accompanied the discovery of the New World certainly didn't help the matter. The Spanish explorer and inspector of mines,

Oviedo, had written a *History of the Indies*. Even having visited the Americas, and seen maize growing there, he thought that it had probably been described by Pliny. He says of Pliny's 'millet of India' – 'I think it is the same as what we call "*mahiz*" in our Indies.'

Fuchs called maize *Frumentum Turcicum* – Turkish corn. He wrote:

> *This grain, like many others, is one of those varieties which have been brought in to us from another place. Moreover, it came into Germany from Greece and Asia, whence it is called 'Turkish corn,' for today the huge mass of Turkey occupies the whole of Asia.*

Maize wasn't the only species whose origin was obscured by this tendency to regard and label anything new and exotic as 'Turkish'. In some cases it's stuck with us all the way to the present day. We still call the American bird *Meleagris gallopavo* a 'turkey'.

In 1570, the penny dropped. The Italian herbalist Matthiolus had read Oviedo, and saw through that confusion between India and the Indies. He was brave enough to suggest that everybody else was wrong – and that maize really had come, across the Atlantic Ocean, from the West Indies. After this, it seems to have been fairly generally accepted that maize was a New World plant – or that at least one variety of it came from the Americas. Some herbalists differentiated between two distinct types of maize. One had yellow and purple kernels, an ear with eight to ten rows, and slender leaves, which was labelled *Frumentum Turcicum*. Another type is described as having some black and brown kernels, and broader leaves, called *Frumentum Indicum*. The implication was that *Indicum* had come from the West Indies, while *Turcicum* or *Asiaticum* came from Asia.

The differences between these two, apparently quite distinct, types of maize suggest an interesting possibility. The first type, *Frumentum Turcicum*, sounds much more like the type of maize now known as a 'Northern Flint'. This variant has very hard kernels, and it doesn't hail from the Caribbean at all. It comes from New England and the Great Plains of North America. Rather than being evidence of fast adaptation in maize brought over from the Caribbean, and spreading from Spain into the rest of Europe, the careful descriptions of

Frumentum Turcicum in those sixteenth-century herbals suggests that there had already been a separate introduction of maize into Europe – this time, from North America.

Another clue appears in the English herbal of John Gerard, first published in 1597. Gerard writes that he has grown maize in his own garden, and that it's called 'Turkie corne' or 'Turkie wheat'. He adds some details about its provenance, and thinks – like many of his contemporaries – that one type came from the 'Turkes Dominions' of Asia. But of the New World sources of this grain, he writes that it comes 'out of America and the Ilands adioyning ... and Virginia and Norembega, when they use to sowe or set it, and to make bread of it'. The mention of both Virginia and Norembega flag up a potential North American source of maize.

Virginia is still familiar to us, as a modern state of the US. It's said to have been named by Sir Walter Raleigh – possibly after his virgin Queen, possibly after an indigenous leader – in 1584, the year in which Raleigh sent his first colonisation and research mission to North America. But Norembega is an odd-sounding name; it starts to appear on sixteenth-century maps, in roughly the area of modern New England. The name also became attached variously to a legendary and fantastically wealthy city – an 'El Dorado' of the north; to a river in Maine; and to a putative Viking settlement – founded by Leif Eriksson, of course. In the nineteenth century, Boston's elite found this last incarnation particularly alluring. They liked the idea that the Vikings had settled New England, and had effectively founded their nation. Eriksson was somehow the acceptable – even the heroic – face of European colonisation. And while Columbus was Catholic, Eriksson was – if not Protestant – then at least Nordic.

There's that possible Viking settlement at L'Anse aux Meadows on Newfoundland – and this island may well have been the Vinland described in the sagas – but this didn't develop into a European colonisation of the eastern seaboard of North America. There's no evidence that the Viking presence in North America extended to New England, and on Newfoundland any early Viking settlements seem to have been very short-lived – and thoroughly extinct by the time the sixteenth-century European explorers arrived.

It seems likely, then, that the 'Norembega' that Gerard is vaguely referring to is not a Viking settlement or a mythical city, but simply the area which would later become known as New England. But the English presence there would only become firmly established in the early seventeenth century, decades after the publication of *Gerard's Herbal*.

In 1606, James I issued a charter to the London and Plymouth Virginia Companies – effectively sponsoring them to form new trading connections and aggressively lay claim to land in North America. In 1607, the English explorer and ex-pirate John Smith, working for the Virginia Company of London, founded James Fort – which would become the first permanent British settlement in North America: Jamestown. John Smith was injured in a battle with Native Americans – being famously (and possibly apocryphally) saved by the chief's daughter, Pocahontas – and returned to England. But he headed back over to North America in 1614, exploring and mapping the area which he would name 'New England'. The *Mayflower* settlers arrived soon after, in 1620, leaving from Plymouth in England, and founding New Plymouth in Massachusetts. This is also recognised as a seminal moment in the history of colonisation, and for some marks the true beginning of the permanent settlement of New England.

So, by the time English settlers had put down these permanent roots in North America, what appears to have been North American – not Mexican – maize had been growing in English gardens for more than two decades. Had someone brought this domesticate over even before the Virginia Companies were granted their Royal Charter? Raleigh's research mission to Virginia in 1584 is clearly too late. But the European presence in North America does go back a little earlier than that. Further north, the English colony in Newfoundland was officially recognised in 1610 – but it had been claimed for the English crown in 1583, by Raleigh's half-brother and fellow adventurer, Humphrey Gilbert.

That's surely still too late for the spread of maize through the gardens of England – just fourteen years before Gerard first published his *Herbal*. But Gilbert wasn't the first European person to set foot on Newfoundland since the Vikings. The European discovery of the island predated Gilbert's voyage by eighty-six years.

Cabot and the Matthew

Hanging in Bristol's Museum and Art Gallery is a huge painting which has entranced me since I was a small child. It was painted by an artist called Ernest Board, who studied art in Bristol and seemed to enjoy historical subjects and large formats. The painting shows a grey-haired man, standing on a quayside, in splendid Medieval get-up, wearing a doublet of red and gold brocade, scarlet leggings and wonderfully long, pointed leather boots. He's gesturing to the ship moored to a post at the quay, while at the same time shaking the hand of an older man in a long, dark robe, who's wearing a mayoral chain of office. Half-hidden between these two is a younger man, with auburn hair, in a red doublet. Behind the mayor in his dark robe – and closer to us – is a bishop, dressed in an embroidered chasuble, his red-gloved hand grasping his golden crook. He's flanked by two small, white-robed acolytes, one carrying a Bible, and the other a candle.

There's a gaggle of other people in the background, all craning to get a better look. In the foreground, a pile of weapons and helmets lies on the cobbles, and a man in a crenellated white hood is picking up an armful of halberds or bills, presumably to load on to the ship. All that we can really see of the ship itself is its prow, but its billowing foresail forms the backdrop for the scene on the quay. Half-hoisted, the sail is painted with a castle and a mast in front – the coat of arms of Bristol. In the distance, we glimpse the Medieval skyline of the city. And, to the right, a tower stands on the horizon. It looks a lot like the Wills Memorial Building, which towers over the city today – but that was only built in 1925. It must be the spireless tower of St Mary Redcliffe. The painting is entitled *The Departure of John and Sebastian Cabot on their First Voyage of Discovery, 1497*. The grey-haired man at the centre of the painting must be John. Standing behind him, in the red doublet, is his son, Sebastian.

Five years after Columbus set sail for the Indies, in a south-westerly direction, under the sponsorship of Ferdinand and Isabella of Spain,

John Cabot left England to sail north-west. He was an Italian by birth, and became a citizen of Venice – so we should really call him Giovanni Caboto, or, for a Venetian twist, Zuan Chabotto. A maritime trader, Cabot (as I'll persist in calling him) worked out of Venice and Valencia, and then turned up in London. He was planning a northerly exploratory voyage across the Atlantic – and this was diplomatically extremely sensitive. A papal bull, or decree, of 1493 had already granted Spain and Portugal exclusive permission to explore the non-European world. Cabot really needed royal support for what would undoubtedly be seen as an incursion into Spanish and Portuguese territory. The Spanish ambassador wrote to Ferdinand and Isabella to explicitly warn them that *'uno como Colon'* – one like Columbus – was in London. But Cabot got the support he needed. Presumably Henry VII didn't see why the Spanish and Portuguese should have it all stitched up, and in 1496 he granted Cabot a licence for exploration. The licence accorded Cabot the right to hold, in the king's name, any land he took possession of, and to have a monopoly over any trade routes he forged. But Cabot still needed financial backing for the voyage. It seems he may have obtained some funds from Italian bankers in London, but also from wealthy Bristol merchants, willing to gamble on this venture. One merchant in particular, who was also a customs officer, has led to the formation of an alluring myth. His name was Richard ap Meryk, also known as Richard Ameryk.

It's generally accepted that the 'Americas' are named after the Italian scholar and explorer Amerigo Vespucci, who voyaged to South America between 1499 and 1502, and realised that those 'West Indies' weren't part of Asia at all – but an entirely new land mass. But what about this Richard Ameryk? His surname has sparked a suggestion that the Americas were in fact named after him. It's a popular explanation, in Bristol at least, but even Ameryk's connection with Cabot is more than a little tenuous. While some have suggested that Ameryk was the principal backer of Cabot's expeditions – and even the owner of the ship in which Cabot set sail, the *Matthew* – there's unfortunately no documentary support for any of these speculations.

Still, the Bristol connection itself is secure. Cabot's charter stipulated that he should sail from this maritime city, which already had some history of Atlantic exploration. A couple of expeditions in the early 1480s had aimed at finding new fishing grounds. But there were also stories of a mythological island called 'Hy-Brasil' that might have stimulated some adventures – and there were even rumours that Bristol sailors had found it. Perhaps some Bristolian really had already discovered North America – even before Columbus made his trip – but we'll probably never know the truth of it.

Cabot set off in 1496, but short supplies and inclement weather forced him back. Undeterred, he got ready to have another go at it in 1497. He left Bristol on 2 May, and reached the other side of the Atlantic on 24 June. Various historians have suggested Nova Scotia, Labrador and Maine as the site of that landfall, but Cape Bonavista, on the east coast of Newfoundland, is thought by many to be the most likely landing place – and it was to there that a replica of Cabot's ship, the *Matthew*, sailed from Bristol in 1997. Some five hundred years earlier, Cabot had been pretty sure that he'd been to the east coast of Asia. Back in England, Bristolians thought he'd probably found the mythical Isle of Brasil.

Cabot returned to the New World for further exploration, but his wanderings are imprecisely recorded. A historian who made some exciting but extraordinary claims about Cabot's adventures, Alwyn Ruddock, died before publishing her research on the subject – and ordered her research notes to be destroyed as soon as possible after her death, which can hardly fail to raise suspicions. But Ruddock asserted that, in 1498, Cabot explored the entire east coast of North America, claiming it for England, and made an incursion into the Spanish territory in the Caribbean.

Amongst the documents that do survive, relating to Cabot's voyages, there's a disappointing dearth of information about the plants and animals he encountered. In stark contrast to the descriptions of Columbus's voyage, no one seems to mention anything that Cabot brought back with him from the New World. After the first voyage, Henry VII gave Cabot a whole ten pounds for his trouble, but commercially the voyages had been a failure. Diplomatically, this venture was a bit of an embarrassment too. While Cabot had been away, Arthur, Prince of Wales, had become betrothed to Catherine of

Aragon – daughter of Ferdinand and Isabella. The marriage was intended to cement an Anglo-Spanish alliance. Better, then, not to tread on Spanish toes, and to sweep that less-than-entirely-successful voyage of exploration under the carpet. The royal marriage went ahead in 1501; Arthur died six months later. But there was still hope for the kingdom, in the form of Arthur's brother. Eight years on, Catherine married that brother – becoming the first wife of Henry VIII.

Still, there was a whole New World out there, and English explorers and pioneers – including John Smith and Henry Gilbert – continued to investigate and lay claim to the northern continent. The names of seventeenth- and eighteenth-century sailors and explorers would become stamped on the map of North America, from Henry Hudson to George Vancouver.

But it was the earlier pioneers who must have introduced the North American varieties of maize to northern Europe – in plenty of time for them to be recorded in *Gerard's Herbal*. John Cabot's son, Sebastian, pictured in that Ernest Board painting, reported that some Native American tribes lived off meat and fish, while others grew maize, squashes and beans. It's impossible to imagine that – in the decades that followed John Cabot's slightly hushed-up discovery of North America – none of these sixteenth-century English explorers brought the northern variant of maize back with them.

And maybe Cabot himself had brought a few grains with him; he would have needed supplies for the return journey, after all. So – imagine Cabot coming home, sailing up the Severn, then the Avon, back into port in August 1497, with not only a head full of new geographical knowledge, but his pockets full of maize kernels. This is a fiction, a figment – as fanciful and romantic as the Board painting – but I love to think that Cabot returned to Bristol and grew sweetcorn in his garden.

Genetic voyages

When history of the more traditional variety, laid out in ink on parchment, vellum and paper, runs out, we can turn to the genetic archive – the precious rolled-up scrolls contained in the nuclei of the cells of the organisms themselves. The nuclear narrative; the chromosome chronicles.

Back in 2003, a group of French plant geneticists published the results of their research into maize genetics. By looking at patterns of difference and similarity among 219 separate samples of maize, from the Americas and Europe, they had hoped to uncover some of that forgotten history. They used a technique which involved cutting up DNA with enzymes, then comparing the lengths of the fragments that had been produced – between different samples. It's essentially the same technique that was developed for forensic purposes, which became known as 'DNA fingerprinting'. It's fairly crude, compared with modern DNA sequencing, but it does reveal patterns of similarities and differences between genomes – and, using it, the French geneticists gained some very clear insights into the saga of maize domestication and globalisation.

They found that maize was wonderfully diverse – much more than had been previously thought. The American populations – especially those from Central America – contained much more variation than the European ones. Maize was clearly, originally and entirely, an American plant – there was no hint of any Asian heritage. Within the Americas, Northern Flint maize, from the higher latitudes of North America, appeared genetically very similar to Chilean varieties. Both these types have long, cylindrical ears and long husk leaves, and hard-as-flint kernels. And genetic similarities between populations of maize on either side of the Atlantic preserved memories of voyages of discovery. Closely related, genetically similar samples of maize would appear as tight clusters in the analysis. The geneticists found that six southern Spanish populations clustered with Caribbean populations – the two were clearly closely related. Presumably the southern Spanish maize varieties were the descendants of the first maize brought back from the New World. But this Spanish maize, quite obviously, had not spread into the rest of Europe. Even Italian maize was different to the Caribbean varieties – it was closest to South American types, from Argentina and Peru. And North European maize was genetically closest to the American Northern Flints. The hints in the herbals – of a separate introduction from North America – were borne out in the DNA of maize growing in northern Europe today. The sixteenth-century German botanist Fuchs was so sure of the Asian, or Turkish, origin of this grain. But his herbal of 1542, the first to contain an illustration of maize, depicted a plant with long ears – with eight to ten rows of kernels – and long husk leaves. It *looks* like a Northern Flint.

Historians have suggested that maize from North America was brought over to Europe in the seventeenth century, but the combination of evidence from genes and the great European herbals pushes the introduction right back into the first half of the sixteenth century – if not slightly earlier. And this isn't at all far-fetched. Archaeological and genetic studies have shown that, by this time, Iroquoian populations were growing maize – as a staple food – across a great sweep of eastern North America – precisely the territory which was being thoroughly explored by English and French pioneers in the sixteenth century.

It's odd that there's such a hole in the historical literature when it comes to maize in the north. But it was such a novel thing, and it seems that words failed the European adventurers. Two explorers, commissioned by King Francois I of France, Giovanni Verrazano and Jacques Cartier, may have referred to maize in rather oblique terms that have been missed in the past. These two were both exploring, and writing about their discoveries, in the 1520s and 1530s. Verrazano writes about an excellent and delectable 'legume' that he tastes when meeting Native Americans living near Chesapeake Bay. Later French texts describe maize as a legume. Cartier, exploring what would become Quebec, describes ceremonial feasts involving 'gros mil' – a term for sorghum, which has surely been appropriated here for maize.

It seems clear that there were ample opportunities for an early introduction of North American varieties of maize into northern Europe – from the end of the fifteenth century into the first half of the sixteenth. More recent genetic analyses strongly suggest that there were indeed multiple introductions of Northern Flints into Europe. Cabot and son, Verrazano and Cartier, are just a few of the pioneers who could have brought Northern Flints back with them. As well as coming back with official voyages of discovery, maize probably hitch-hiked with unofficial Atlantic fishing expeditions. And in contrast to the tropical Caribbean maize, the North American varieties were already adapted to temperate climates – they would have thrived immediately in central and northern Europe.

The genetic story of maize plays out in a similar way in eastern Asia. Maize in tropical latitudes, from Indonesia to China, is closest to Mexican maize. But this time, history supplies the details – the Portuguese introduced maize into south-east Asia as early as 1496, and another wave of maize arrived with the Spanish colonisation of the Philippines in the sixteenth century. The genetic map of maize in Africa is complicated, with early introductions of South American maize to the west coast by Portuguese colonisers in the sixteenth century. This history echoes in the African names for maize – *mielie* or *mealies* – which derive from the Portuguese word for maize, *milho*. Later, from the nineteenth century onwards, varieties from the southern half of North America, known as 'Southern Dents', were introduced to eastern and southern Africa. Up in the north-west corner of Africa, there's evidence of Caribbean ancestry – just as in southern Spain. That Caribbean genetic signal is also scattered across western Asia, from Nepal to Afghanistan. Linguistic and historical clues support the role of Turkish, Arab and other Muslim traders in the spread of maize from the Middle East, by sea and by land – from the Red Sea and the Persian Gulf out into the Arabian Sea and eastwards to the Bay of Bengal; along the Silk Road and through the Himalayas.

But it's the DNA of maize from the middling latitudes of its new homes around the world that's most fascinating. In the north of Spain and the south of France, European maize is equally related to North American and Caribbean types. It looks like hybridisation created the perfect in-betweener – as early as the seventeenth century. Strains of maize which had diverged away from each other, adapting to different environments, in the Americas, were brought back together in the foothills of the Pyrenees.

The dissemination of maize across the world was astonishingly fast. Genetic analysis and molecular dating suggests that maize was domesticated around 9,000 years ago in the Americas. It stayed in this region for 8,500 years, going global in just the last 500 years. But in fact its spread was even faster than this implies – the documentary evidence shows that maize had spread right across Eurasia, from Spain to China, in just six decades after Columbus first brought it over from the Caribbean. In some ways, this spread and adoption seems quite extraordinary – these were regions of the world where agriculture had been practised for millennia, and there were already well-established fields of wheat and rice to provide populations with staple foods. The historical records show that farmers didn't immediately swap their traditional crops for this new grain. Instead, maize was often grown on marginal land, by impoverished farmers trying to eke out a living in relatively barren areas. It was considered a food of the poor – and yet, once it had a foothold in the Old World, the global future of maize was assured. Its sheer variety, and ability to grow in such a wide range of environments, meant that – as soon as it crossed the Atlantic – it was poised to spread throughout the world.

American origin

Back in the Americas, genetic studies have been crucial, not only in estimating the timing of maize domestication, but in tracking down the identity of the wild progenitor, establishing how many times maize was domesticated, and *where* this happened. Maize is a subspecies – *Zea mays mays* – and there are three other subspecies within the same species – all of which are wild, and known more colloquially as teosinte: a name that comes from the Aztec language of Guatemala. The Aztecs venerated maize, in the forms of the goddess Chicomecoatl and the god Cinteotl.

The three teosintes – *Zea mays huehuetenangensis, mexicana* and *parviglumis* – grow wild in Guatemala and Mexico. Although the teosintes look quite distinct from their domesticated cousin, maize hybridises freely with all of them. If we imagine evolution as a branching tree, it seems likely that one of these cousins will be closer to maize than the others, and may even represent the surviving wild descendants of the original population that was also domesticated.

Analysis of enzymes in maize and the teosintes had suggested that one of the wild subspecies was indeed more similar to maize than the others. And in 2002, this was confirmed by a large genetic study. Having tested 264 samples in total – of maize and the three teosintes – the geneticists found that Mexican annual teosinte *Zea mays parviglumis* was closest to the domesticate.

As the study contained so much data on American maize populations – 193 of the 264 samples were from maize – it was also possible to construct a phylogeny, a family tree, for this domesticate. All the maize lineages – from the temperate-adapted Northern Flints to the tropical types in Colombia, Venezuela and the Caribbean – tracked back and coalesced, converging on a single stem. So maize was domesticated just once. Or at least, if it was domesticated several times, only one, branching lineage has survived to the present. The stem of the phylogenetic tree was rooted in Mexico. But pinning down the place where the domestication first started was tricky. The most primitive form of domesticated maize on that phylogenetic tree grows in the highlands of Mexico. But the closest wild relative is a lowland plant: it's the *Zea mays parviglumis* of the Balsas River Basin of central Mexico, or Balsas teosinte.

By the time this genetic information emerged, the earliest evidence of maize in the archaeological record – in the form of whole cobs – came from the Mexican highlands, dating to 6,200 years ago. So it seemed that, either Balsas teosinte had been carried up into the mountains to be planted, or it was first domesticated down in the valleys, spreading to higher altitudes later.

Over thousands of years, climate and environments will have changed quite a bit, and species will have shifted accordingly. But, given the new genetic data and the identification of the closest wild relative to maize, archaeologists believed it was still worth having a good look down in the Balsas Valley. And so they began to scour the area for traces of ancient cultivation and domestication. What they needed was something which would clearly distinguish the wild from the domestic.

When it starts growing, teosinte can be difficult to distinguish from its domesticated cousin, making it a vexatious weed in maize fields. But when it matures, it looks quite different. Each teosinte plant is bush-like, with branching stalks – whereas maize grows with a single, tall stalk. Teosinte ears are small and simple, with a staggered row of about a dozen kernels attached to a central rachis. Maize cobs are huge in comparison, crammed with hundreds of kernels. Teosinte kernels are small, and each contained in a hard case; maize kernels are large and naked. And just like wild wheat, wild teosinte ears shatter at maturity, whereas maize kernels stay firmly attached to a non-shattering rachis. Geneticists have been able to pinpoint just a handful of genes which have undergone mutations to produce the differences in branching, kernel size, fruitcases and seed shattering between teosinte and maize.

This is all very well, but down in the tropical lowlands, preservation of plant remains is pathetic at best – the archaeologists had no hope of finding whole plants, whole cobs, or even intact kernels. Instead, they turned their attention to much smaller components of plants – phytoliths and starch granules. Phytoliths are silica-rich and very resistant to degradation, meaning that they stick around, even in tropical places, for an incredibly long time. Both the phytoliths and starch granules of teosinte are – very usefully – characteristically distinct compared with those of maize.

The first evidence of these microscopic traces of early maize were discovered in lake sediments in the Balsas River Valley. The archaeologists followed up by excavating four prehistoric rock shelters in the region – and one of them, the Xihuatoxtla Shelter, yielded precious, early evidence of maize. Stone tools from the cave – in a layer dating to 8,700 years ago – contained diagnostic maize starch granules tucked into cracks and crevices. Maize phytoliths were also found on the stone tools, as well as being scattered throughout samples of sediment from inside the rock shelter.

The phytoliths provided further clues as to how the ancient Mexicans were using maize. It's been suggested in the past that maize may have been cultivated, first and foremost, for its stalks. The hard fruitcases of ripe teosinte kernels would have made them unpalatable, whereas the sugary pith of the stalk could have been eaten or even used to make a fermented drink – a sort of teosinte rum. Phytoliths are different in the stalk and cobs of maize, and the archaeologists working on the samples from Xihuatoxtla found plenty of cob phytoliths but none from stalks. It seems that it was the grain that the early cultivators were most interested in – at least at this site. And the kernels appeared to have already undergone a genetic change associated with domestication, shedding their hard fruitcases – as no phytoliths from such cases were found. Other sites in Panama, dating to around 6,000 to 7,000 years ago (4000 to 5000 BCE), have presented a similar picture – of the use of cobs, not stalks. It's still possible that hunter-gatherers may have used the sugary stalks of teosinte more than its grains, and switched to a focus on grains later, when the plant had already begun to develop domesticated features. But perhaps the difficulty of processing teosinte kernels has been overplayed. They can be made edible by soaking and grinding, and some Mexican farmers still use teosinte seeds to feed their livestock.

This discovery of early maize, in the seasonal tropical forest of the Mexican lowlands, is important. It significantly predates – by two and a half millennia – the previous evidence which was used to argue for an origin of domestication of this crop in the highlands. It also makes a lot more sense – Balsas teosinte, the closest relative of maize, grows naturally in the lowlands, not up in the mountains.

Yet, after all this sleuthing, there's still a big, juicy question that remains. After 1493, this home-grown American crop rapidly spread all over the world, into myriad environments, getting a toehold even in some of the world's most inhospitable landscapes. The global success of maize depended on its large portfolio of variation – but how had it developed all that astonishing variety, coming from a single origin in the lowlands of south-western Mexico?

Extraordinary and conspicuous diversity

In his book *The Variation of Animals and Plants Under Domestication*, published nine years after his *Origin of Species*, in 1868, Darwin wrote about the American origin, antiquity and wonderful variety of maize:

> *Zea mays ... is undoubtedly of American origin, and was grown by the aborigines throughout the continent from New England to Chili. Its cultivation must have been extremely ancient ... I found on the*

coast of Peru heads of maize, together with eighteen species of recent
sea-shell, embedded in a beach which had been upraised at least 85
feet above the level of the sea. In accordance with this ancient
cultivation, numerous American varieties have arisen ...

Darwin didn't know about that close relationship between annual
Mexican teosinte, particularly that in the Balsas Valley, and maize.
'The aboriginal form [of maize],' he wrote, 'has not as yet been
discovered in the wild state.' But then he gives an account of a young
Native American man who told the French botanist Auguste de
Saint-Hilaire about a curiously maize-like plant – but with husked
seeds – that 'grew wild in the humid forests of his native land'.

Darwin was impressed and intrigued by the 'extraordinary and
conspicuous manner' in which maize varied. He believed that the
dissimilarities between varieties had arisen as the crop spread into
northern latitudes, developing an 'inherited acclimatisation' to
different environments. He writes about the experiments of the
botanist Johann Metzger, who tried growing various American
varieties of maize in Germany – with remarkable results.

Metzger grew some plants from seeds obtained from a tropical
region in America. And this is how Darwin described the outcome:

During the first year the plants were twelve feet high, and a few seeds
were perfected; the lower seeds in the ear kept true to their proper
form, but the upper seeds became slightly changed. In the second
generation the plants were from nine to ten feet in height, and ripened
their seed better; the depression on the outer side of the seed had
almost disappeared, and the original beautiful white colour had
become duskier. Some of the seeds had even become yellow, and in
their now rounded form they approached common European maize.
In the third generation nearly all resemblance to the original and very
distinct American parent-form was lost. In the sixth generation this
maize perfectly resembled a European variety.

This is such an astonishingly quick transformation. It seems much too
quick to be down to a genetic change in the plants. It sounds more

like physiological adaptation, or – if you can bear even more technical jargon – phenotypic plasticity. This concept relates to the latent potential – which is still governed by genes – for organisms to adjust, during a lifetime, to particular environments. Adult organisms usually have a limited ability to adapt physiologically or anatomically in this way. But organisms nurtured from birth, or grown from seed, in a different environment to their parents can end up looking quite dissimilar, and functioning differently too.

Darwin's writing is brilliant in so many ways. He builds arguments beautifully, and he illustrates big ideas with carefully described, often personally experienced, details – like those ancient maize cobs which he found in the raised beach in Peru, 85 feet above sea level. Sometimes he's laying out his argument, and providing evidence to support a particular theory. But at other times, you can almost feel the whirring of his mental cogs. He's endlessly inquisitive and excited by new pieces of information that reach him. With Metzger's tropical American maize grown in Germany, Darwin's much less surprised by changes to the stem, and the time it took for seeds to ripen, than he is by the transformation of the seeds themselves. He writes: 'It is a much more surprising fact that the seeds should have undergone so rapid and great a change.' But then he almost argues with himself, introducing the dialectic into his own monologue: 'As ... flowers, with their product the seed, are formed by the metamorphosis of the stem and leaves, any modification in these latter organs would be apt to extend, through correlation, to the organs of fructification.'

In other words, flowers – and their seeds – develop out of the tissues of stem and leaves. So if stem and leaves are being modified by climate, perhaps it's not so surprising after all that seeds change so much as well. In this passage, Darwin gets very close to understanding something that we can now appreciate from a genetic perspective. Separate parts of an organism are not always controlled by separate genes – far from it. The relationship between DNA, on the one hand, and the form and function of a whole organism, on the other, is much more complicated than that. A change in a particular gene can have widespread effects throughout the body of an organism – whether that's a human, a dog, or a maize plant.

With his discussion about the astonishing changes observed in tropical maize after just a few generations of growing in that less favourable climate in Germany, Darwin is also getting very close to that, much more recently articulated, idea of phenotypic plasticity. What we now know is that this doesn't require a change in the DNA itself – what might be called a 'true' evolutionary change. It just requires a modification to the way the organism reads, or expresses, its DNA. Even without genetic mutations, phenotypic plasticity can be a source of extraordinary novelty. And yet so much research into the transformation of wild species into domesticated ones focuses purely on genetic mutations, sometimes forgetting just how much the phenotype can vary, without a change to the underlying DNA code. Metzger's tropical maize, transplanted into a temperate climate, is a fantastic example of just how malleable the phenotype can be. And one recent study uncovered an even more surprising degree of plasticity than Metzger had demonstrated with his American maize.

Dolores Piperno is an archaeobotanist at the Smithsonian Museum in Washington DC. She led the investigation that found maize phytoliths in the Xihuatoxtla Shelter in the Balsas Valley. But as well as looking for ancient traces of long-dead plants, her research also involves doing experiments with their living counterparts. She led a team from the Smithsonian Tropical Research Institute in Panama, which – between 2009 and 2012 – set about examining just how important a factor phenotypic plasticity might have been in the variation produced in maize, as it became domesticated. They took the wild ancestor of maize, *Zea mays parviglumis*, and grew it in glasshouses under two sets of climatic conditions. One climate replicated that of the end of the Ice Age, between 16,000 to 11,000 years ago. The other was a control chamber, replicating the modern climate. As the plants in each chamber grew, the results were astonishing.

In the modern control chamber, all the plants looked just like wild teosinte – with lots of branches, sprouting both tassels and female ears. The kernels of the ears ripened in a staggered fashion, rather than all at once. The late-Ice Age chamber was somewhat different. Most of the plants looked like teosinte, but some – about one in five – looked very much like maize. These plants developed a single stem, rather than lots of branches. Attached directly to the main stem were female flowers which developed into ears of corn where all the kernels ripened at the same time.

It's always been a bit of a mystery as to why teosinte seemed like an attractive candidate for cultivation to those early farmers. But if some teosinte plants – back at the end of the Ice Age – looked more like maize does today – with ears close to the stem and easy to harvest, and seeds ripening all at once – then perhaps it's not so strange after all.

Something even more intriguing happened when the researchers took seeds from just the maize-like plants grown under glacial conditions, and grew them in a climate matching that after the Ice Age, just into the Holocene, 10,000 years ago. Half of the plants from those seeds *still* looked like maize rather than teosinte. This means that early cultivators could have very quickly ended up with plants which mostly had that desirable, maize-like phenotype. We know that genetic changes also took place as maize became domesticated, but it seems that phenotypic plasticity is an important part of the story. The impressive plasticity of maize may represent an adaptation to variability – it suggests that its ancestors were exposed to fluctuating conditions and were more successful if they could adjust quickly to novel growing environments. We can no longer overlook this phenomenon – phenotypic plasticity – if we really want to understand how plants (and animals) became domesticated – and the important role that environment and ecology play today.

And so, changing its form in response to climate and to selection by its human cultivators, maize began to spread from its homeland in the tropical forests of Mexico – up into the highlands, and into more northerly and more southerly latitudes – as the craze for agriculture took hold. The gradual spread of maize through the Americas allowed it to adapt to different environments – crucially becoming, not only a lowland plant, but a highland one; not only a tropical plant, but a temperate one.

Phenotypic plasticity and new genetic mutations are two important sources of novelty, helping to produce the 'extraordinary and conspicuous' diversity of maize. But there was something else that seems to have contributed to its amazing ability to adapt to new environments – a little help from its wild relatives. As early maize spread from the lowlands to the highlands of Mexico, it hybridised with the mountainous subspecies of teosinte, *Zea mays mexicana*. Genetic studies have shown that up to around 20 per cent of the genome of highland maize comes from *mexicana*. Just like domesticated barley, picking up its drought resistance from wild strains growing in the Syrian Desert, maize was making the most of local, genetic 'knowledge' as it spread – by hybridising with its wild relatives.

Maize appears to have migrated from Mexico via separate highland and lowland routes, into Guatemala, and on, further south. It had reached northern South America by 7,500 years ago. By 4,700 years ago, maize was growing in lowland Brazil, and by 4,000 years ago, it was in the Andes. From northern South America, maize spread northwards to Trinidad and Tobago, and the other islands of the Caribbean. The spread of maize to North America was much slower – beginning in the south-west corner just over 2,000 years ago, but then spreading right up to the north-east, into what is now Canada, in perhaps just a few centuries. And as maize spread, it kept changing.

By the time of European contact with the Americas, a huge range of varieties of maize had developed, growing everywhere from Mexico to north-east America, from the coasts of the Caribbean and the valleys of Brazil, up into the heights of the Andes. In all its different forms, it was already a highly adapted and highly variable domesticate – primed and ready to spread rapidly across the globe – as soon as Columbus planted his foot on that beach.

5

POTATOES

Solanum tuberosum

The coarse boot nestled on the lug, the shaft
Against the inside knee was levered firmly.
He rooted out tall tops, buried the bright edge deep
To scatter new potatoes that we picked,
Loving their cool hardness in our hands.

Seamus Heaney, 'Digging'

Ancient potatoes

A fragment of grey, crumpled, thin, leathery material – so small it would almost fit on a fingertip. Quite uninspiring. If you found it in your back garden, you'd think it was merely recent detritus. Something that had escaped from the compost heap, perhaps. (Something as unremarkable as a bit of chipped stone pushed out from the burrow of a lobster.) And yet, this is a very precious piece of archaeological evidence.

This little black piece of organic stuff came from an archaeological site that was excavated during the 1980s in southern Chile, called Monte Verde. It's one of the oldest, securely dated sites of human habitation in the Americas, North or South, at around 14,600 years old. It's almost contemporary with the Natufian sites in the Levant, but the big difference is that modern humans had lived in the Near East for tens of thousands of years before this. At Monte Verde, they were still relative newcomers.

I visited Monte Verde in 2008 – with geologist Mario Piño, who had dug at the site. We arrived at this incredibly important place to find a field, with a few sheep grazing on the mossy banks of the fast-flowing Chinchihuapi Creek. We were so far from England, but I could have been hiking in the Lake District – it was such a familiar rural idyll. And without Mario's expert help, I would have had immense difficulty locating the precise location of the site – the archaeology had all been covered over, and had completely merged back into the landscape. In fact, I probably wouldn't even have known it was there.

'The site, like so many others, was discovered by chance,' Mario told me. 'Local villagers were widening the creek. When they were removing the sediment, cutting the curves, they found huge bones, and kept them. Two university students who were travelling round the place took the bones to Valdivia.'

It was lucky they did. The huge bones turned out to be from Ice Age animals – that went extinct around 11,000 years ago. The find prompted scientists from Valdivia University to investigate further. What seemed at first to be a purely palaeontological site, containing the remains of Pleistocene animals, became even more interesting when the researchers began to find stone tools and other relics. People had clearly been here too – a very long time ago.

The wet, peaty soil at the site meant that organic material was extremely well preserved. Things that would have quickly rotted away at most sites had survived here, trapped in a time capsule. The archaeologists began to find the remains of wooden stakes, stuck into the ground, and it soon became clear that these were outlining the frame of a building – some kind of hut. And it was big – about

20 metres long. In the soil around the stakes there were dark fragments of tough organic material: the animal hide that had been used to cover the long cabin. The archaeologists also found evidence of fire pits or hearths – full of charcoal – both inside and outside the building. The preservation was astonishing. There was even a child's footprint, perfectly preserved in the mud. About 30 metres away, they uncovered evidence of a smaller hut, full of animal and plant remains, including butchered mastodon bones and chewed-up and spat-out lumps of seaweed.

The site appears to have been abandoned and then fairly rapidly buried – it seems likely that the area had become very boggy and, after the humans left, reedbeds quickly took over. The peat built up, sealing the archaeology and preserving all those precious organic remains – forgotten, until the villagers decided to widen the creek.

The preservation of organic matter at the site gave the archaeologists an unprecedented opportunity to look at the whole range of animals and plants that formed the diet of the hunter-gatherers who lived there. The Monte Verdians had eaten the meat of now-extinct animals – including elephant-like gomphotheres and ancient llamas – and a huge variety of plants: forty-six different species in all. The plants included four species of edible seaweed – some present as those chewed-up cuds – which may also have been used for medicinal purposes. And amongst the plant remains were those tiny, unprepossessing, leathery scraps – the crumpled remnants of the skins of ancient, wild potatoes: *Solanum maglia*. Nine fragments were found in all, from small hearths or food pits inside the huts. Analysis of starch grains still sticking to their inner surfaces confirmed the species. These are still the earliest remains of potatoes ever discovered, in association with humans – our ancestors had already developed a taste for the humble spud some 14,600 years ago. Wooden digging sticks, perfect for unearthing those potatoes, were also found at the site.

'We found food from all four seasons,' said Mario. So it seemed that this place was more than just a seasonal camp – it had been used all year round. This is intriguing too, as we tend to assume that people at this time were all quite nomadic, making temporary camps, striking them and moving on. The slightly later, Mesolithic site of Star Carr challenged this assumption back in England. Monte Verde presents us with the same reality check in South America. We shouldn't look for a one-size-fits-all for any period in the past – or indeed, the present – and we shouldn't underestimate just how sophisticated our ancestors were. In some places, it would have made sense to stay mobile. In others, the conditions and resources of a particular region meant that settling down in one place was a perfectly viable way of life. Human behaviour changes to suit local ecology.

The site of Monte Verde has sparked some controversy, due to its early date. During the twentieth century, a prevailing hypothesis held that the earliest inhabitants of the Americas arrived, in the north, around 13,000 years ago, carrying with them a particular stone toolkit known as 'Clovis', after the site in New Mexico where characteristic stone points were discovered in the 1930s. Monte Verde is clearly too early to fit with that model.

By 1997, lead archaeologist Tom Dillehay was so fed up with the criticisms – that the dating of Monte Verde couldn't possibly be right – that he invited a group of eminent colleagues to visit the site, to see the artefacts for themselves and make up their own minds about it. They all agreed that the site was indeed archaeological, and that there was no reason to doubt the pre-Clovis radiocarbon dates.

Monte Verde is now just one of several 'pre-Clovis' sites providing solid evidence of people inhabiting the Americas much earlier than the 'Clovis-first' hypothesis would allow. The consensus view is still that the first colonisers arrived in the north, moving across the Bering land bridge from north-east Asia. There are a couple of very early sites in northern Yukon, indicating human presence at those high latitudes, which date to before the peak of the last Ice Age, 20,000 years ago. But a massive ice sheet effectively sealed off most of North America. Colonisation of the rest of the continent, and then South America, had to wait until the ice began to melt. Pre-Clovis sites in North and South America show that the colonisation must have happened fairly soon after the last glacial maximum, perhaps around 17,000 years ago. Although much of North America was still under a massive ice sheet at this time, environmental analyses have shown that the northern Pacific coast would have thawed out enough for people to have taken this route into the Americas. Then they spread southwards, giving them just about enough time to have reached Chile, by 14,600 years ago.

Just how long was it until the early hunter-gatherers in South America discovered those little parcels of deliciousness hiding in the soil? My guess is: not long at all.

Digging for tubers might seem like an extraordinarily inventive way of obtaining food. Picking fruits and nuts off trees, even collecting seaweed off rocks on the beach – those all seem like fairly obvious approaches to foraging. On the other hand, sharpening yourself a digging stick and going poking around for hidden, underground morsels: this seems on the face of it to be either deeply strange or desperate behaviour – or alternatively, a stroke of genius.

But our ancestors have been carrying on in this way for not only thousands but, potentially, *millions* of years.

Buried treasure

Our closest living animal relatives are chimpanzees and gorillas. Both of these forest-living apes prefer eating ripe fruit – but when this food is scarce, they fall back on leaves and the pithy insides of stems. It seems likely that, around 6–7 million years ago, the common ancestor of humans and chimpanzees would have relied on a similar diet. But then the ancestors of humans and chimpanzees diverged. The apes that belong to our own branch of the family tree of life on the planet are known as hominins – characterised by habitually walking around on two legs, and possessing increasingly large brains compared with their forebears. We're the only living representatives of a once bushy hominin sprig on the tree of life. We now know of around twenty hominin species – all of them extinct, except us. When early hominins start to appear in the fossil record, they not only show skeletal adaptations to walking on two legs, their teeth have changed too – they have larger molars with much thicker enamel than their predecessors. In other primates, tooth shape and size seems to relate less to the preferred, everyday diet and more to the types of foods that animals resort to when times are hard. This suggests that the change to hominin teeth may also reflect a change in fallback foods. This was a time when the great, dense forests of Africa were starting to break up. The landscape was becoming more diverse – and it seems that our ancestors were beginning to exploit these more open environments.

There are some obvious differences between savannah and forest ecosystems, but underground there's an important, hidden contrast. Savannahs contain many more plants with 'underground storage organs' – such as rhizomes, corms, bulbs and tubers. Comparing a modern savannah in northern Tanzania with the rainforest of the Central African Republic, ecologists have found an enormous difference in the density of tubers and other underground storage organs: 40,000 kilograms in every square kilometre of the savannah, compared with a meagre 100 kilograms per square kilometre in the forest. Were our ancestors tapping into this particular rich resource, under the expanding grasslands of Africa? The prize for digging out these tubers, and the like, was a parcel of energy – but a tough one. It may not have been the food of choice, but it could make a big difference in desperate times. Those larger, well-armoured teeth of our early ancestors could represent an adaptation to this new fallback food.

Contemporary foragers make good use of roots, tubers and bulbs. I've been fortunate enough to have seen for myself how one group of modern hunter-gatherers – the Hadza – take advantage of this particular type of food. In 2010, I went on an expedition to meet a Hadza group in a remote part of Tanzania – with anthropologist Alyssa Crittenden.

Flying in to Kilimanjaro Airport, I then set off in a four-by-four. The first half of the journey – about three hours – was easy enough, on tarmac roads, passing through small villages. But then we suddenly took a left turn on to a dirt track, and for most of the next three hours I was being thrown around in the Land Cruiser as the driver, Petro, drove capably along rutted tracks, down into sandy stream beds and up steep banks, until we reached the edge of Lake Eyasi – a vast salt flat with little sign of any water. We dropped down into the lake – and got stuck on the rim, at an awkward angle. There was nothing we could do – this vehicle was well and truly immobilised.

It was late and dusk was falling fast. We didn't fancy spending the night in the Land Cruiser, so we called through to the advance party of the team, who had already arrived and set up camp. They came to rescue us in another Land Cruiser – winching out our vehicle.

We weren't far from the camp, and when we arrived there I met Alyssa, the anthropologist who had been studying and living with an indigenous, hunter-gatherer group for years. Our tented safari camp had been pitched near the Hadza camp, under the trees. I thought everyone might have already gone to bed, but Alyssa said that the Hadza were excited about meeting me. So in the gathering darkness, I went with Alyssa and was introduced to a group of around twenty people, shaking hands and saying '*Mtana*'. The women wore dresses and kangas of bright, printed material, and a few had beaded head-bands. Some of the men were wearing T-shirts and shorts, others wore loincloths, and necklaces of black, red and white beads. Everyone's hair was cut very short. I gave out the small gifts that Alyssa had asked me to bring: little bags of beads for the women and steel nails for the men, which would be hammered into arrowheads. These people met me with such warmth and openness – as a friend of a friend.

In the few days I spent with the Hadza, I felt as though I learnt a huge amount about their way of life – although it was really just a glimpse. I was extremely lucky to have Alyssa there as my guide; her depth of knowledge was extraordinary. I watched Hadzabe men and boys mending their bows and arrows and then setting off on a hunt. I also observed – from a safe distance – a man braving the stings of angry bees to collect honey from a hive hanging in a tree. On return to the camp, he was mobbed by women and children, all wanting a piece of the honeycomb. I talked to the Hadzabe women – through two layers of translation – about having children and childcare. And I accompanied the women when they left camp to go on foraging expeditions into the bush. They had a specific target in mind – and that was tubers.

On one such foray, Alyssa and I set off with the women. The children all came along too – infants carried on their mums' chests in simple cloth slings, toddlers jogging to keep up, the older kids running and skipping. We walked south from the camp for just under a mile, pausing to eat berries on the way. Eventually we all stopped at a dense thicket. The women and children then disappeared *inside* the bushes, digging around the roots of climbers for tubers. The tubers, which were called '*ekwa*', were not at all what I had expected – more like swollen roots than the potatoes in my vegetable bed at home. I crawled inside the thicket with a woman called Nabile, who was heavily pregnant – but she wasn't letting that stop her. She showed me how she dug with a pointed digging stick, and I had a go – it was a useful tool. Breaking up the hard soil and working the *ekwa* loose with the point, it could then be scooped out with our hands. Nabile would occasionally pause in her digging and take out a knife to sharpen the tip of the stick. We were soon down to the roots of these bushes. Freeing up a section of root from surrounding soil, Nabile would then use the knife again, to cut a piece out – and she'd immediately begin to eat it. These pieces of tuber were about 20 centimetres long and 3 centimetres thick. She'd rip the outer, bark-like layer off with her teeth, then use a knife to make a shallow cut, allowing her to tear off a strip of the root, which she folded up and chewed. She offered me some too. The taste was a pleasant surprise – the first crunch reminded me of biting into a stick of celery, though the taste was completely different. It was quite fibrous, but nutty and moist.

As well as eating some of the roots raw, where they'd dug them up, the women had each collected a good haul, into their cloth shoulder bags, to take back to the camp. Once there, they got the fires going again and roasted the roots over the embers. I was given a piece to try. Now the skin peeled off very easily, and the flesh inside was much softer – and delicious. It tasted a bit like roasted chestnut.

Spending just a little time with the Hadza opened my eyes to their way of life – and my own – in a manner that's difficult to communicate. I came back with a new perspective on my own culture, from how we manage to balance work and family life, to what we eat. It's all too easy to look at other cultures, in the present and the past, with rose-tinted spectacles, but I still felt that we – in the 'Western' world, could learn a lot from these traditional approaches to life. It may not all have been rosy, but with a focus on family and community, there were no 'jobs' – and no unemployment. Everyone had a role to play. And children were part of it. There was no suggestion that having children could be detrimental to a woman's status in society.

Back to food – I was surprised to see how prized honey was. Men returning with honey were welcomed back more eagerly than those returning with meat. The desire for sweetness is always there – it only becomes a problem when sugar is as cheap and accessible as it is in places like Britain. And as for a varied diet – the Hadza had access to a much broader range of foods than I'd naively assumed, but it was really striking to see just how important roots were in that diet.

Root and tubers are actually fairly low-quality foods – they have nowhere near the amount of energy that's packed into fruit and seeds, meat and honey. But they are *dependable*. Anthropologists asked Hadza about food preferences, and found that honey – the most energy-dense food in nature – came out on top. Tubers consistently ranked lowest. Meat, berries and baobab fruit come somewhere in between. But despite the low ranking of tubers, they're also the food-stuff that makes up the bulk of the Hadza diet – precisely because they can depend on them. Weighing the separate food types that came back to camp, the anthropologists found that proportions changed from season to season, and also varied for groups in different regions. Tubers appeared to be both a staple food, eaten all year round, and a fallback food – relied on even more when other foods were scarce.

The fact that most hunter-gatherers in tropical latitudes dig up and eat roots or tubers suggests that humans have probably been doing this for a very long time – perhaps as long as modern humans have been on the planet. That's more than 300,000 years. But the thick enamel and large teeth of early hominins suggests that this behaviour has even more ancient – forgive me – roots. A simple digging stick could have given our ancestors a vital survival advantage on the plains of Africa. But all of this seems very speculative. These are good hypotheses, certainly, but we need to test them. Is it possible to find any firmer evidence of our forebears eating tubers?

The answer is – to a certain extent, yes. Advances in the analysis of fossils now enable us not only to make interpretations based on the size and shape of bones, but to look closely at their chemical compo-sition. As all the tissues of your body are, ultimately, built out of the molecules that you ingest, it's possible to find clues about ancient diet entombed in those fossil bones.

Specific chemical elements exist in subtly different forms, called isotopes. Some of these are stable, whereas others are unstable, radioactive versions. There are three naturally occurring forms of carbon. There's the unstable, radioactive carbon-14, which is rare, but extremely useful to archaeologists as it's used for radiocarbon dating. Most carbon in the world exists in the form of carbon-12, which has six neutrons and six protons in its nucleus. But there's also a slightly heavier – and still stable – version, with an extra neutron, called carbon-13.

When plants photosynthesise, they use sunlight energy to drive a reaction which captures carbon dioxide from the atmosphere, and eventually builds that carbon into brand-new sugar molecules. There are a few types of photosynthesis, each using slightly different chemical pathways. Trees and bushes tend to use a form of photosynthesis which includes the formation of a molecule with three carbon atoms as an early step. Ingeniously, plant scientists decided to call such plants 'C3 plants'. Then there are plants like some grasses and sedges, which do photosynthesis slightly differently, creating a four-carbon molecule. You can see where this is going. They're called 'C4 plants'.

Not only is the C4 pathway more efficient in its use of water molecules – making it a useful adaptation in more arid environments – it also means that the plant grabs more of the slightly heavier stable isotope of carbon-13. So C4 plants are relatively carbon-13-enriched. If an animal eats a lot of C4 plants – including, for instance, the roots and corms of sedges – the animal itself ends up being carbon-13-enriched too – even its bones.

Anthropologists have used this difference between C3 and C4 plants to good ends. Chimpanzees' diets are dominated by leafy C3 plants – their bones do not end up carbon-13-enriched. Our early hominin ancestors, some 4.5 million years ago, seemed to eat a similar, C3 plant-based diet. Between 4 and 1 million years ago, the climate was fluctuating, but the landscapes our ancestors inhabited were – on average – becoming drier and grassier. By about 3.5 million years ago, we know that our ancestors were eating a mix of C3 and C4 plants – and it might be that the C4 contribution came from starch-rich roots and tubers. Eating those hidden but ubiquitous foods could have helped ancient populations to expand and thrive in new habitats – including variable and unpredictable environments.

Then, by 2.5 million years ago, there's a split. Some hominins – which also happened to have very robust teeth and jaws – were eating mainly C4 plants (perhaps grass blades, seeds, sedge corms, depending on the season). At around the same time, other hominins, including the earliest members of our own genus, *Homo*, continued to consume a mixed C3–C4 diet.

Although it's often been argued that the advent of regular meat-eating provided the energy needed for our ancestors to evolve bigger brains, some researchers have recently suggested that plant foods – and in particular, starchy plant foods, like tubers – have been rather overlooked. Two key developments – one cultural, and one genetic – would have hugely helped to unlock the energy bound up in starch. That cultural development was cooking; the genetic one was the multiplication of a gene that produces an enzyme in saliva, to break down starch. We know that this gene multiplication happened some time after a million years ago. Salivary amylase works so much better on cooked than on raw starch, so it may be that the increase in the copies of this gene came hot on the heels of the adoption of cooking. There are archaeological suggestions of humans using fire as early as 1.6 million years ago, and definite evidence of hearths by 780,000 years ago. Together, cooking and plenty of salivary amylase may have provided the energy – in the form of ready-to-use glucose – for the enlarging human brain. And of course, a similar adaptation to eating starchy foods developed in dogs. Although dogs don't produce salivary amylase, they do produce this starch-busting enzyme in their pancreas – and many of them have multiple copies of the pancreatic amylase gene.

We know that our ancestors have been making and using stone tools for more than 3 million years. Those tools may have been used for processing both meat and plant foods. What's really lacking in the archaeological record is any organic remains. So we have no idea when our ancestors started to use digging sticks. But as soon as they invented this simple tool, they would have had access to that buried treasure – that dependable resource which would become something of a staple, and a fallback food, for so many hunter-gatherers.

What we can say with some certainty is that, by the time people were living at Monte Verde, their ancestors already had a long history of using digging sticks and eating roots and tubers. Eating wild potatoes was just the latest, local manifestation of that ancient type of behaviour.

But when – and where – did potatoes become transformed from a gathered, wild food to a cultivated, domesticated species?

The Cave of Three Windows and the unsolved riddle

The Chilean wild potato, *Solanum maglia*, is a pretty plant with white flowers and small, purplish tubers, less than 4 centimetres in diameter, which likes to grow in damp ravines and around the edges of bogs, close to sea level, near the coast of central Chile. The species name comes from its name in the language of the indigenous Mapuche people of central Chile: *malla*. Darwin saw these plants in 1835, during his voyage on board the *Beagle*. He knew that the explorer Alexander Humboldt had written about these wild plants, and believed them to be ancestral to the domesticated potato. Darwin noted in his journal:

> The wild potato grows on these islands in great abundance, on the sandy, shelly soil near the sea-beach. The tallest plant was four feet in height. The tubers were generally small, but I found one, of an oval shape, two inches in diameter: they resembled in every respect, and had the same smell as English potatoes; but when boiled they shrunk much, and were watery and insipid, without any bitter taste. They are undoubtedly here indigenous ...

The domesticated potato, *Solanum tuberosum*, cultivated all over Chile and beyond, is very similar to its wild cousin. So similar, in fact, that even Darwin misidentified a *Solanum tuberosum* specimen he'd collected as *Solanum maglia*. But with the aid of microscopy, identification becomes much easier – it was the starch grains clinging to the inner side of the fragments of potato skin at Monte Verde that proved them to be the remains of wild *Solanum maglia* tubers.

The archaeologists who dug Monte Verde wanted to taste the wild potatoes for themselves. They acquired a tuber, boiled it for half an hour, and ate it. It was a brave thing to do. Some researchers had suggested that wild potatoes would be too bitter to eat. They tend to contain relatively high levels of glycoalkaloids – such as solanine – which are part of the potato's natural defence mechanism against infection and insects – and, it could be argued, against being eaten by humans. Glycoalkaloids impart a bitter taste to potatoes, and at high levels they're toxic. It was thought that wild potatoes could contain such high levels of these compounds that they'd still be poisonous, even after cooking.

But – like Darwin – the archaeologists not only survived the experiment, they didn't detect any bitter taste to this mini-potato. Although some wild potatoes, further north in the central Andes, do produce bitter tubers, the wild Chilean potato seems to be perfectly pleasant to eat. And the archaeologists also reported that local inhabitants of central Chile happily eat wild potatoes today.

But is *Solanum maglia* the ancestor of the domesticated potatoes we now eat? This is – or, at least, has been – a highly contentious question. As with many species, the question started out as that familiar one: was this a case of a single centre of domestication – or were there multiple origins?

There are hundreds of types of potatoes, and botanists have argued about how to organise these into varieties and species. Some are between-species hybrids, making this task even harder. Classifications have resolved the different types into as many as 235 species, but the latest analysis – including genetic data – suggests that all potatoes can actually be grouped into 107 wild species and four cultivated species.

Some of the most ancient varieties, or landraces, of potatoes are grown high up in the Andes – up to 3,500 metres above sea level – from western Venezuela to northern Argentina, and down in lowland south-central Chile. These landraces can be grouped into four species.

One of these species, *Solanum tuberosum*, contains within it two clear, separate cultivars or subspecies – an Andean group and a Chilean group.

In the early twentieth century, Russian botanists proposed that there had been two main centres of potato domestication – high up on the Peruvian and Bolivian plateau, near Lake Titicaca, and low down, in southern Chile. But then English botanists came up with a different model: a single origin of potatoes up in the Andes, and then an expansion of these domesticated potatoes southwards, to coastal Chile – adapting to the local conditions there. This seemed to match up well with the evidence – there were many more wild species from which *Solanum tuberosum* could have arisen, up in the Andes, compared with Chile.

The earliest evidence of domesticated potato does come from the Andes – from a cave called Cueva Tres Ventanas (The Cave of Three Windows) in the Peruvian highlands, nearly 4,000 metres above sea level. The cave contains the oldest mummies in the world – dating to between 8,000 and 10,000 years ago, but the potato remains come from a younger layer, dating to around 6,000 years ago. And experiments have shown that Andean-type potatoes could be fairly easily transformed into something that looked like a Chilean type. For a while, then, it looked like a single origin of domesticated potatoes, high in the Andes, was the most likely scenario.

But by the 1990s, another hypothesis had emerged – that the Chilean type developed as a hybrid of the Andean type with local, wild species in Chile. That wild species was suggested to be *Solanum maglia* – the same species of wild potato that was eaten at Monte Verde. But there's such a enormous number of wild species, and potato genetics is hugely convoluted. And yet, finally, some form of clarity does seem to be emerging from the chaos. It seems that both the Russian and the English botanists were partially right. The latest archaeological and genetic evidence suggests that a wild potato species was first domesticated somewhere around Lake Titicaca, in the high Andes, between 8,000 and 4,000 years ago – around the same time that the llama was domesticated. But genetic studies also provide support for the hybrid origin of the Chilean potato cultivar, meaning that, as the original Andean domesticate spread, it hybridised with other, wild species. So, more than one wild species contributed to the gene pool of the first domesticates, and the simple question of origins (too simple for complex, interwoven, entangled biology) becomes more nuanced. Are we looking at multiple, independent centres of domestication, and separate lineages which are later brought together by interbreeding in some cultivars? Or are we looking at a single origin, in a discrete area, with a subsequent spread and interbreeding with other species? From a genetic point of view, it probably doesn't matter so much. However it happened, genes from the lowlands and the highlands were brought together in the Chilean cultivar. But from a human perspective, this is a pertinent question, because it becomes about culture and innovation. Did the idea of growing potatoes emerge and take hold just once? Did this idea gradually spread down into the foothills of the Andes and thence to the coastal plains of Chile? Or, once hunter-gatherers started eating potatoes, was it almost inevitable that some wild species would become domesticated, and that this then happened in at least two locations, possibly more? A single origin may be more likely, but it seems to me that we don't quite have the tools or the evidence to answer that question yet. There's still more work to be done before this particular mystery is solved.

The potato goddess, the mountain and the ocean

Wherever domestication first took off, it transformed the wild potato into something much more useful to humans. The most impressive difference between wild and domestic potatoes is in the size of the tubers and the length of the runners – the thin, horizontal stems that are sent out to sprout new plants. Wild potatoes have very long runners, which allow those new plants to propagate a good distance away from the parent; and they have small tubers. Domestication has trimmed the runners much shorter, and promoted larger tubers – both features which make the potato plant less fit in the wild, but much easier to harvest. It's like the tough-rachis characteristic in wheat – an abject disadvantage for a wild plant, but a boon for one that's teamed up with humans. Domesticated potatoes also contain much less of those glycoalkaloids that can make some wild potatoes so bitter, and even poisonous.

Potatoes gradually became more and more important to Peruvian societies – and Andean civilisations rose. By the first millennium CE, potatoes had become embedded in society – they were a crucial, staple crop. The Inca Empire, which emerged in the twelfth century CE – stretching from Ecuador to Santiago – was fuelled by this subterranean commodity. The Inca even had a – slightly lumpy – potato goddess, called Axomama. And they grew so many varieties of potato that they needed to invent imaginative names to differentiate them – from the sinuous *Katari Papa*, the 'snake potato', to the difficult-to-peel *Cachan huacachi*, the 'potato that makes the daughter-in-law weep'.

A couple of millennia before laughing Martians helped to make dehydrated instant mash popular in Britain, the ancient Andeans had hit upon this method of preservation. It helped that they lived, essentially, in a freezer – at least, once the sun set. By night, potatoes were laid out on the ground, to freeze. During the day, they'd thaw out and would be trodden on, to squeeze out water. Then they'd be left out to freeze again. After three or four days and nights the potatoes were transformed into chuño – freeze-dried potatoes. As well as dehydrating the tubers, this processing would also drive out glycoalkaloids from the chuño, making it less bitter than fresh potatoes. While domestication would have involved the selection of the most palatable potatoes – and this probably started before cultivation – some potatoes remained a little too bitter. Another way of reducing bitterness was to eat the potatoes with clay, which binds to the glycoalkaloids. Around Lake Titicaca today, there are still some Aymara people who eat their potatoes this way. Perhaps even more importantly, making chuño transformed potatoes into a form that could be stored for extended periods, sometimes years. While the elite amongst agricultural societies in the Fertile Crescent grew wealthy by amassing stores of wheat and herds of cattle, the Inca chiefs grew rich and fat on their stores of dried potatoes. Chuño became a currency in its own right – the peasants paid their taxes in it, while labourers and mercenaries were paid in it.

By the time of European contact with the Americas, domesticated potatoes were widely grown in western South America, from the Andean altiplano down the Chilean lowlands. And as the Spanish moved in to South America in a more aggressive way, they came to understand the value of chuño. High in the Bolivian Andes, 4,000 metres above sea level, they found a mountain so full of silver that it became known as Cerro Rico, the 'Rich Mountain'. The Inca had mined it for centuries. For the Spanish, it was an unmissable opportunity. The treasure that Columbus had dreamt of was here for the taking. As the mines poured forth silver, a town grew up at the foot of the mountain: Potosi. It became the site of the Spanish colonial mint, and in the sixteenth century 60 per cent of the world's silver came from there. At first the Spanish sent Native Americans down the mines – some conscripted, some there to earn a wage – but the work was dangerous and life-limiting. As the indigenous labour force dwindled in the seventeenth century, the Spanish mine-owners switched to bringing in African slaves, tens of thousands of them. And they were fed on chuño. Translating the energy stored in potatoes into an unimagined bounty of silver, the Spanish flooded the European markets with the precious metal.

The Andean silver arriving in Europe fulfilled the promise of the New World – fabulous riches really were to be found there. But in the depths of the Rich Mountain, the price was paid for dearly in human life and misery. And the suffering didn't stop there. The influx of silver into Europe fed inflation and destabilised economies. Meanwhile, the food that had fuelled the mines was also making its way over to Europe. Potatoes were coming to the Old World.

But which of the closely related subspecies of *Solanum tuberosum* – the high Andean or the low Chilean – was first introduced to Europe? Unsurprisingly, there have been proponents for both. These two cultivars vary quite subtly in their physical characteristics – the Chilean varieties have wider leaflets than the Andean. But it's the adaptations to geography and climate which are most important. More crucial than altitude and temperature, it's adaptation to latitude that's critical here.

Potatoes from the Andes, from modern-day Colombia, had evolved in a place relatively close to the equator, where they'd become used to twelve hours of daylight. For these potatoes, moving to a more seasonal latitude would have been challenging. It was not so much the short days of winter that would have been a problem, but long summer days. Too much daylight inhibits the formation of tubers. But Chilean cultivars – growing further from the equator – would already have adapted to relatively long hours of daylight in the summer.

Plant physiologists have elucidated the factors controlling tuberisation. The leaves of the potato plant detect sunlight and length of day, and send out chemical signals which affect the development of roots and tubers. Some of the essential chemical signals have been identified. There's a phenomenon in molecular biology (and astronomy) where the first compounds (or celestial bodies) to be discovered are often given quite genial names. Then the imagination of the scientist is stretched too far, and subsequent molecules (and stars) are assigned a string of letters – usually an acronym harking back to the longer names of related compounds – and numbers. So tuberisation involves a panoply of players, from phytochrome B, gibberellins and jasmonate to miR172, POTH1 and StSP6A. You may be relieved that I have no intention of spending the rest of this chapter describing the entire process of tuberisation and our present understanding of its molecular basis. (You may also be disappointed, and I'm sorry – but it's not that type of book.) Suffice it to say, the physiology of tuberisation is impressively complicated. So we have a familiar conundrum – how do you alter part, or many parts, of this machinery without derailing the process entirely? And what are the chances of a random mutation popping up to do just the thing which would prove to be of benefit as potatoes spread further into temperate latitudes?

Even with everything we now know about how evolution works, this still seems to be something of a philosophical sticking point. But it's not an insuperable hurdle – it can't be, because we know potatoes have done it, somehow. We know that small changes to certain genes can alter the role of particular key players in biochemical pathways. Genes with such important, fundamental roles are often called 'master control genes'. The proteins they encode, known as regulatory factors, act as molecular switches – turning other genes on or off, or more subtly controlling how strongly a gene is expressed. So it's possible for a minor alteration in one gene –encoding one of these important molecular switches – to have significant and widespread effects. Even though evolution, at a genetic level, works its magic through tiny changes, some of those tiny changes can have profound, far-reaching consequences for the phenotype – the structure and function – of an organism: evolution can make sudden jumps.

There's a good candidate for just such an important molecular switch – or regulatory factor – in potato tuberisation, meaning that a tiny alteration really could result in a pronounced physiological change. The variation that already exists within populations represents an important part of the solution too. A species is not one organism, one genome. It is the sum of all of its parts, and those parts vary. As potato-farming spread southwards, into latitudes with longer summer days, some potatoes would have been better able to produce tubers than others. In a more temperate climate, those variants would have had the edge. Natural selection would weed out the rest.

Given these adaptations to latitude, it seems likely that Chilean potatoes would have stood a better chance in Europe than their more equatorial counterparts from the northern Andes. In 1929, Russian botanists proposed precisely this origin for the European potato. But British researchers were sure that the original European potatoes hailed from the Andes. Historical records suggested that potatoes arrived in Europe at a time when the Spanish were barely established in Chile, whilst having conquered countries around the northern Andes – Colombia, Ecuador, Bolivia and Peru – half a century earlier.

Many botanists believed that the balance of probability lay in one particular direction. For the last sixty to seventy years, the prevailing hypothesis has followed that British suggestion – that European potatoes descended from northern Andean stock. This hypothesis seemed strengthened by the fact that ancient varieties of potatoes in the Canary Islands and in India looked as though they harked back to northern Andean roots.

Then the geneticists got involved, and – as they tend to do so often – set the cat among the pigeons. The Canary Island potatoes turned out to contain a mixture of Chilean and northern Andean heritage. The Indian potatoes were quite clearly of Chilean origin.

Their interest aroused, the geneticists moved on to look at mainland European potatoes – undertaking genetic analyses on historical samples from herbarium collections dating to between 1700 and 1910. The eighteenth-century potatoes of mainland Europe proved to be of largely northern Andean heritage. They must have had to adjust quickly to long summer days. Perhaps this was due to rapid adaptation – brought about by something like a novel mutation occurring in a particular, master molecular switch, producing widespread effects. In fact, such a mutation need not have been entirely novel – long-day variants may have already existed in those newly imported Andean potatoes; we know that this characteristic pops up from time to time in those varieties. Maybe the adjustment to temperate latitudes wasn't quite as challenging as had once been thought.

But that wasn't the end of the story. In samples from 1811 onwards, the geneticists found evidence of Chilean ancestry in European potatoes. Some previous researchers had suggested that Chilean varieties had been introduced after blight epidemics had swept through the earlier, northern Andean stock, starting in 1845. There was always a problem with this hypothesis, as Chilean potatoes are not especially resistant to blight. Nevertheless – for whatever reason – Chilean varieties had clearly been introduced to Europe in the nineteenth century – and they quickly became very successful. Although the northern Andean varieties were the first to be established in Europe, Chilean potatoes seem to have had an inbuilt advantage, perhaps drawing on their deep history of growing somewhere where summer days were long – it's their DNA which predominates in the European varieties we grow today.

Carmelite friars and a bouquet of potato flowers

As for how potatoes arrived in Europe in the first place, you may be fairly sure that Columbus brought them back from the New World, just as he did with maize. But it's not true. Although Columbus and other adventurers did ship many foods back to Europe in those early days of contact with the Americas, the potato was not among them. And this is because it was cultivated on the *western* side of South

America – from the mountains down to the Chilean lowlands – and the Spanish did not reach the High Andes until the 1530s, some forty-odd years after Columbus's first explorations across the Atlantic. The first written report of potatoes came from Spanish explorers in 1536, who found them growing in the Magdalena Valley of Colombia.

As a further complication, there is no historical record of the first arrival of potatoes in Europe. Whoever received them on this side of the Atlantic obviously did not deem them particularly worthy of note. Or perhaps they did, and their exhilarating account has somehow been lost to history. There's also a linguistic complication: sweet potatoes (*Ipomoea batatas*) are *batata* in Spanish, while *Solanum tuberosum* is *patata*. However, the first published reference to what really does seem to be a potato appears, in Spanish literature, in 1552. Soon after, there are records of potatoes in the Canary Isles. The first mention of potatoes actually appearing in Europe – as an import, rather than a crop – dates to 1567, when there's a record of them being shipped from Gran Canaria to Antwerp.

(Forgive me a little detour here. There is apparently a heated debate about who really invented French fries: the Belgians or the French. Both countries claim priority, with the Belgians blaming 'French gastronomic hegemony' and geographically challenged American soldiers for the naming of this particular delicacy. The first documentary evidence of such treatment of potatoes is apparently Belgian, and dates back to the late seventeenth century, according to unverified journalistic sources. On the other hand, the very first documentary evidence of potatoes reaching the European mainland is that consignment making its way to Antwerp. We have no way of knowing what the Belgians actually did with those potatoes. But I like to think that someone in Antwerp, 450 years ago, may just have invented what almost became a national dish. Until the French stole it, anyway.)

Just six years after that first mention of potatoes in Europe, there's fairly firm evidence of cultivation, in Spain. The 1573 accounts of the Carmelite Hospital de la Sangre in Seville describe potatoes being bought in the last quarter of that year. This strongly suggests that the potatoes were locally grown, seasonal vegetables. It also points to potatoes being grown in the autumn – in a growing season with short days, then, which would have suited Andean varieties admirably. Very much like Caribbean maize, potatoes hailing from tropical latitudes (regardless of altitude) in the Americas seem to have settled in relatively easily to southern, Mediterranean Europe.

Once the potato had gained a toehold in Spain, it spread quickly to Italy – introduced there by Carmelite friars. And then, again rather like maize, this exotic vegetable began to disperse through the botanical gardens of Europe, making its appearance in herbals written in the late sixteenth century. The Swiss botanist Gaspard Bauhin gave it its Latin name, *Solanum tuberosum* – 'the soil-bound swelling'. The English botanist John Gerard – who thought that one variety of maize came from Turkey – was equally confused about the origin of potatoes. He was so sure the potato had come from Virginia that he named it *Battata virginiana*. And thus, he sowed the seeds of the legend that Sir Walter Raleigh brought potatoes to England from his colony in the New World. Another legend, that Sir Francis Drake transported potatoes from Virginia to England, is equally devoid of any basis in fact.

Having been introduced and spread within Europe via seemingly elite networks that involved the Catholic Church, potatoes appear to have been enthusiastically adopted by the peasantry of Italy, who by the beginning of the seventeenth century were eating them alongside turnips and carrots and also feeding them to pigs. Meanwhile, potatoes were spreading eastwards too, reaching China in the same century. Within the Americas, the northwards expansion of the Spanish Empire saw the introduction of potatoes to the west coast of North America. Potatoes also made their way back from Europe, across the Atlantic, with British traders and migrants. By 1685, William Penn was able to report that potatoes grew well in Pennsylvania.

But the craze for potatoes was slow to spread north in Europe. The reasons for the late adoption of this vegetable seem to include some deep-rooted but rather odd superstitions. Potatoes, perhaps because of their odd, misshapen tubers, like deformed limbs, were linked to leprosy. The fact that potatoes were not mentioned in the Bible was also a source of suspicion. The similarity of potatoes to deadly nightshade caused consternation, and perhaps this wasn't an entirely undue worry. Once potatoes turn green, and start sprouting, the levels of solanine in them are enough to be quite toxic. That's why it's so important to keep potatoes in the dark. Learning how to store potatoes safely would have been crucial to avoid being poisoned by them. Other concerns about potatoes included possible flatulence and an increase in lust, hopefully not at the same time. And beyond that, there was a general aversion to eating potatoes in many countries where they were first adopted as a crop destined to feed animals. When, in 1770, a boatload of potatoes was sent as famine relief to the starving inhabitants of Naples, they rejected it.

Beyond the taboos and superstitions, there may have been a more prosaic reason for the slow acceptance of the potato in northern Europe. From a purely functional point of view, it was tricky to slot potatoes into the three-year crop-rotation system that had been practised across Europe since Roman times. It was awkward for individual farmers to make changes in one of their strips, within a larger field that they shared with the other farmers in the village.

Eventually, the cultural barriers to potato expansion came, if not crashing, then at least crumbling down. A curious mix of religion and politics finally conspired to propel potatoes, from southern Europe, northwards and eastwards. In the late seventeenth century, the Huguenots and other Protestant groups were expelled from France, taking their expertise – in areas as diverse as silversmithing, obstetrics and potato cultivation – with them, wherever they went. In the mid-eighteenth century, the fallout from the Seven Years' War demonstrated another advantage of the potato: lurking underground, this crop – unlike cereals – could survive in fields that had been burned and trampled. When Antoine-Augustin Parmentier, a pharmacist in the French Army, was caught by the Prussians, he was fed potatoes in his cell. Rather than baulking at this treatment – after all, he was only familiar with potatoes as fodder for livestock – he was impressed by the nutritional value of his prison fare. When he returned to France in 1763, he became a vociferous proponent of potatoes. He hosted potato-based dinners for the great and good, and gave bouquets of potato flowers to Louis XVI and Marie Antoinette. But it was a run of poor harvests, revolution and famine which eventually secured the place of this humble tuber in French cuisine. Today, Parmentier's pioneering spirit is remembered in the name of many French dishes, all of them involving potatoes in some form or another. His grave in Paris is ringed with the plants he loved so much.

Helped by Parmentier in France, and other champions including Frederick the Great in Germany, and Catherine the Great in Russia, potatoes broke out of the monasteries and botanic gardens, into the fields of the northern European plain. Potatoes began to replace traditional staples and fallback foods such as turnips and rutabaga, providing a real alternative to the sometimes risky reliance on cereals that had gone before. Famines still hit from time to time, but less frequently now that there was another staple to fall back on. Together with that other American import, maize, potatoes helped to support an astonishing growth in the European population, which almost doubled in the hundred years between 1750 and 1850, from 140 million to 270 million. Potatoes – once the fuel of the Inca Empire – now provided a huge economic boost to the countries of central and northern Europe – providing energy for a growing population, and underpinning urbanisation and industrialisation. While the steam-powered machines of the Industrial Revolution were fed on coal, its workforce was powered by – cheap, dependable, plentiful – potatoes. The balance of political power in Europe began to shift, from the warmer, sunny countries of the south to the colder, greyer states of the north. The factors behind the rise of the European superpowers in the eighteenth and nineteenth centuries are many and complex, but somewhere in there, darkly lurking, is the potato. And it was there in the crises of the twentieth century – forming an important provision for armies. Second World War military rations included that old Andean trick – dehydrated potatoes.

Potatoes played their part in history, as empires rose and fell, and battles were lost and won. But potatoes were changing too. The nineteenth and early twentieth century saw the creation of a great range of new cultivars, as potatoes and other domesticates became subjected to intense selective breeding. While potatoes had once allowed the Spanish slave-drivers to unlock the silver from Potosí, this New World crop was finally becoming respected as a treasure in its own right. Potato breeders became fabulously wealthy. One new variety, created at the beginning of the twentieth century, was even named Eldorado. But this treasure from the Americas also brought with it a curse.

Feast and famine

Potatoes became another staple food in Europe, complementing grain, and helping to improve food security – up to a point. The problem arose when countries started to depend on the crop too heavily – and it owed much to the way that this crop is propagated. When potatoes failed, they failed badly.

If you want to plant potatoes in your garden, you can buy a bag of seed potatoes. The name is, of course, completely misleading. These are potatoes, certainly, but they are not seeds. The plants that emerge from these small potatoes are clones of their parents, which have been grown in carefully controlled conditions to keep lines pure and minimise interbreeding between separate cultivars. Potatoes are flowering plants – and quite pretty ones at that, with their five-petalled lilac flowers – and the whole point of flowers is sexual reproduction. As insects visit the blooms to take what they need, in the form of nectar, they bring pollen from other plants. Pollen is the plant equivalent of sperm: it contains a half-set of plant chromosomes – male DNA from another plant, or even the same plant. The important thing about this DNA is that it's been mixed up a bit as the pollen is made; the same thing happens as the ovum is formed. The germ cells that give rise to the gametes – the pollen and ovum – contain pairs of chromosomes. Within each pair, the chromosomes swap genes with each other, during meiosis – the special type of cell division which forms the gametes. (This is the point at which duplications can happen – remember the multiplying amylase genes in dogs.) A gene on one chromosome may differ a little from the equivalent gene on the other chromosome. Just one of each chromosome pair passes into the pollen grain or ovum, with a selection of gene variants pulled from one or other of the original pair – so this is already something new, and different to the chromosomes of the parent.

When pollen and ovum combine, the chromosomes derived from each parent pair up – and a completely novel combination of gene variants, or alleles, is created. Sexual reproduction is all about creating novelty and variation. But potatoes also reproduce, quite naturally, by asexual reproduction. In fact, this is precisely what the tubers are *for*, from an evolutionary perspective. Not for human (or any other animal) consumption, but for creating new versions of themselves.

While gathering seeds from potatoes to plant next year's crop may be possible, it's not the most obvious way to create another generation. It's far easier to save a few small potatoes to do just that. Using seed also introduces an element of uncertainty about next year's plants – sexual reproduction guarantees a level of variation which is far from welcome if you're trying to grow plants with particular characteristics. Using 'seed' potatoes eliminates that uncertainty – in fact, the potatoes you're planting aren't really a new generation at all. They are the identical twins of the plants from which you took those potatoes. This is asexual reproduction: the new crop is a clone of the old.

This may sound like a good idea – if you have a crop with particular, desirable characteristics, you surely want to hang on to those features. But eliminating variation is a dangerous game to play. The fact that so many plants and animals reproduce sexually is important – it *works*. Creating variation with every new generation provides the possibility of new variants which will have an advantage – especially if the environment changes. So generating variation is nature's way of future-proofing species. The environment is more than the physical situation in which a plant or animal lives – it's biological too: it involves all the other biological entities that may interact with that particular organism. Many of those entities pose a threat: they may be viruses, bacteria, fungi, other plants, animals. And those potential enemies are always evolving better ways to attack, better ways to avoid any defences which have evolved in the organism under threat. It's nothing less than an arms race – and if the defender fails to keep up, its fate is clearly marked.

If you grow potatoes from seed potatoes, and keep some potatoes from that crop to plant again, and keep on doing that – you have trapped those potatoes in suspended evolution. You may be able to protect your potatoes from other, potentially damaging or competitive plants – a little weeding should sort that out. You may be able to protect your precious plants from animals which would like to chomp on leaves or tubers (though beetles can be extremely difficult to guard against). But the most sinister and pernicious threat comes from pathogens so small that they are invisible to the human naked eye: the viruses, bacteria and fungi. Make no mistake, the pathogens – the evildoers – will not be holding back. They'll be evolving newly powerful and mischievous ways to get at your potatoes. And ultimately, they'll win. If there's decent variation amongst your potatoes, there's a chance that some will be endowed with resistance, and survive the onslaught. If there's very little variation, then the pathogen could be utterly devastating. It could wipe out a whole crop. It could wipe out a whole country's worth of crops. And that's precisely what happened in Ireland in the 1840s.

While other north-west European countries were slow to adopt potato farming, Ireland broke the mould. When English immigrants introduced the crop to Ireland in 1640, potatoes were enthusiastically embraced. The Irish farmers found in the potato something that they could cultivate for themselves – on the poorest plots, while their more fertile fields were dedicated to growing grain for absentee landowners in England. The potatoes introduced to Ireland in the mid seventeenth century were probably still essentially Andean varieties. It may seem odd that they settled in so easily, in such a northerly latitude. But the Irish climate was so mild – with September just as warm as June – that potatoes could be grown well into autumn. A potato whose ancestors had grown used to short days, near the equator, would tuberise just as happily, close to the equinox, in temperate Ireland.

By the nineteenth century, Irish farmers were still exporting the vast majority of their grain to England, while they and their families depended on potatoes – almost to the exclusion of anything else. But in this verdant, well-watered isle, the farmers had no way of storing their harvest. They grew them, and ate them, and grew them again. And the genetic diversity amongst the crops was astonishingly narrow. Farmers grew just one type of potato, the Lumper. It was a nationwide experiment in clonal monoculture – and it was doomed.

In the summer of 1845, a fungus called *Phytophthora infestans* reached Irish shores. Its spores may have arrived on a ship from the Americas. The Irish potato crops had no resistance to this novel pathogen. The spectre spread through them with astonishing speed, its spores carried on the wind from field to field; leaves and stems blackened, and underground the tubers turned to pulpy mush; the air was full of the stench of putrefaction. The blight struck again in 1846, and again in 1848. It tore through potato crops right across Europe, but it was in Ireland that the effects were most apocalyptic.

With astonishing, brutal disregard for the plight of the farmers, grain was still shipped out to England. Social injustice compounded the biological tragedy. The Irish farmers and their families had no other staple crops to rely on, and starvation, typhus and cholera stalked the land. The tragedy set in train by the blight became known as *An Gorta Mór*, the Great Hunger – or the Irish Potato Famine. People left Ireland in droves; the famine prompted a huge exodus of Irish refugees, heading west across the Atlantic. Those who made it to North America were the lucky ones. Back in Ireland, in just three years, a million people died. The population of Ireland today is still smaller than it was before the famine and the mass emigration – around 5 million people compared with over 8 million back in the 1840s.

The terrible tragedy holds important lessons for us today. Controlling the characteristics of the plants we grow and the animals we keep for food seems so desirable. It allows us to manage supply and demand, to plan ahead. But it comes at a price – a potentially devastating price – if it means that we prevent domesticated species from evolving, especially where pathogens are concerned.

It seems paradoxical that we've managed to create such great vulnerability when the entire development of agriculture might be viewed as an exercise in managing risk. The lifestyle of the hunter-gatherer seems so precarious compared to that of the farmer: one relies on nature providing, the other controls the harvest and stores any food left over as insurance, ready for hard times – but can also translate any surplus into wealth and power. But it seems as though our control of nature may be less complete – and even much more illusory – than we'd like to think. We've ended up trying to pin biology down, to stop it changing, when the fundamental way of nature is *to change*. By limiting evolution in domesticated species, we can make them exquisitely vulnerable.

And surely hunter-gatherers have something to teach us about flexibility. They may use tubers as fallback foods, but they try very hard not to rely on just a few sources of sustenance. I'm not suggesting we all adopt hunting and gathering. The global population is far too large for that to be an option. Agriculture has supported a huge expansion in the human population, but at the same time, in some ways, we've become trapped by this cultural development. This seems like a paradox too. With a whole world of plants and animals to choose from, we've narrowed our options right down. On the face of it, the Columbian Exchange created new diversity on both sides of the Atlantic – but globally, we've come to rely on a relatively small range of plants and animals. And within those domesticated species, diversity can be precipitously, dangerously low. The genetic diversity among domesticated potatoes far away from their Andean homeland is paltry today.

A single Andean farmer may grow more than a dozen distinct varieties of potato. These varieties are very diverse in appearance, from the colours and shapes of the tubers and flowers, to their growth patterns. Each cultivar evolves to be suited to a subtly different ecological niche up in the mountains, where the conditions vary dramatically over short distances. In contrast, the thrust of industrialised agriculture is to focus on fewer and fewer varieties, committing huge areas to monoculture. Not just monoculture – but clonal monoculture. We're breeding organisms that are inherently fragile.

Michael Pollan, who occupies his own ecological niche somewhere between nature writing and environmental philosophy, wrote that 'to Western eyes, the [Andean] farms look patchy and chaotic ... offering none of the familiar Apollonian satisfactions of an explicitly ordered landscape.' And yet, these farms, where the various types of cultivated potatoes can breed fairly freely with wild neighbours, and where variety acts as insurance against pests and droughts, increasing the chances that at least some cultivars will survive, seem to provide a more robust solution than industrial monoculture. However knowingly they've done it, the Andean farmers have been successful in cultivating and preserving genetic diversity in their crops.

Farmers have recognised the problems of inbreeding for centuries, possibly even millennia. Producing a population of animals or plants with very little variation may satisfy cultural mores and supermarket requirements, but leaves those organisms dangerously susceptible to disease. Rare breeds and cultivars represent a precious library of much wider genetic diversity, making it incredibly important to preserve those varieties – at the very least, in the case of plants, to collect and store their seeds. Maintaining large, diverse genetic libraries – out in the landscape as well as archives like seed banks – may be our best chance of future-proofing our domesticates. Somewhere in that library is the potential for resistance against diseases which may not even have emerged as a threat yet, as well as the capacity for creating other new, desirable traits.

But there is another way of injecting novel, protective – or otherwise useful – genetic traits into a domestic strain. Selective breeding works, but it's slow and doesn't always produce the desired result. For centuries, it's all we've had, and it has of course produced impressive changes in domestic plants and livestock. But our ability to change organisms to suit our needs has been transformed by new technology – where we alter the genes themselves. Genetically modified plants can be designed to be resistant to a particular pathogen. In the mid 1990s, farmers in North America were experimenting with growing potatoes called 'New Leaf', which had been genetically engineered to produce their own toxin, to fend off Colorado beetle infestation. These potatoes were 'transgenic': a gene from another organism – in this case, a bacterium – had been introduced into the plant's genome.

Genetic modification may prove to be a useful weapon in our armoury – but it certainly doesn't replace the need to preserve genetic diversity. It will never end the arms race between crops and pathogens – evolution will not stand still. It's also – still – a controversial technology. It can introduce novelty into the genetic code with perhaps unpredictable effects. But it can also involve transferring genetic information from one species to another, transgressing that species boundary. And it breaks another 'biological rule'. With selective breeding, the farmer effectively chooses between available genetic variants. He or she does not *create* the variant to start with. As Darwin wrote in *On the Origin of Species*, 'Man does not actually produce variability.' But with genetic engineering – that's precisely what we are doing. Concerns have been raised about the potential – though unknown – long-term effects of breaking the species boundary. There are worries about the escape of novel genes into wild plants. And then there are suspicions about the motivations of the large corporations that have been pushing this technology through.

In the end, New Leaf potatoes never really took off. These GM potatoes were expensive, they required complicated crop rotations to lessen the likelihood of resistance developing amongst the beetles, and then a new, effective pesticide appeared on the market. It was market forces rather than ethical objections that put an end to this particular experiment, after less than ten years.

But perhaps we shouldn't be turning our backs on GM just yet. There's another way of employing genetic technology to produce some of the specific attributes we want to feature in our domesticates – and that's by looking for desirable versions of a particular gene which already exist in that species' genetic repertoire, and spreading that gene through the breeding population. This time it's not about moving a gene across species boundaries, but instead short-circuiting the traditional practice of selective breeding. I wanted to understand how this 'gene editing' worked in practice, so I made an appointment to visit the Roslin Institute in Edinburgh – to meet the geneticists there, and their flock.

6

CHICKENS

Gallus gallus domesticus

A hen is only an egg's way of making another egg.

Samuel Butler

The chicken of tomorrow

Today, chickens outnumber humans on the planet by at least three to one at any time. They are the most common birds on earth, and around 60 billion of them are raised and slaughtered each year to feed our hunger for their flesh. Chickens have become the most important agricultural animal on the planet. But it wasn't always this way. In fact, the chicken's rise to global domination happened very recently, and very fast. And it all started with an American competition launched in 1945 – to find the chicken of the future.

The idea behind the competition was to refocus chicken breeders on meat, not just eggs, and to find the plumpest chicken in the United

States of America. The sponsors of the competition, leading poultry retailers, A&P Foodstores, made a film about it in 1948, ingeniously entitled *Chicken-of-Tomorrow*.

The film starts with a close-up on a crate of fluffy chicks, while an oboe plays a plaintive air. Then the music fades, and the picture changes so that now we see two women in white shirts, fondling the cute little cheeping chicks and throwing them from one crate into another. 'Did you know that poultry is the nation's third largest agricultural crop, a 3-billion-dollar business?' intones the American infomercial-style voiceover. The informative script is read by no less than the film-maker and broadcaster Lowell Thomas – the voice of 20th Century Fox's newsreels until 1952.

Then we're looking at more women, this time transferring eggs into a rack. 'Breeders have achieved great results in boosting the egg output of the average hen. Today's hen averages 154 eggs per year. Some birds produce over 300 annually.' This sounds good, but it's not good enough. 'But with this emphasis on egg production, poultry meat has been more or less a by-product of the industry,' continues the voiceover. Now we see two men in white coats, inspecting very skinny, dead chickens and then hanging them back up by their feet on hooks. The poultry industry got a boost in wartime, we're told, filling a gap in the market produced by a shortage and rationing of red meat. Poultry leaders were worried about maintaining the demand after the war, so A&P Foodstores – originally the Great Atlantic & Pacific Tea Company – stepped into the breach, sponsoring a national competition. They were quite clear about what they wanted from farmers and breeders: 'A broad-breasted bird with bigger drumsticks, plumper thighs and layers of white meat.' They'd even made a wax model of what they wanted the future chicken to look like. Essentially, they wanted a chicken that looked more like a turkey.

The film goes on to describe the state contests, though it's hard to take this seriously – the accompanying jaunty music, 'The Liberty Bell' march, was later hijacked by Monty Python, of course. Then we're on to chick embryology, with footage of chicken embryos developing inside their eggs, a section of eggshell removed at each stage to provide a window on their development.

Back to racks of untampered-with eggs, as we're told that all
entries in the national final were incubated, hatched and raised under
identical conditions. We see five men in suits from the poultry industry
inspecting the chicks, approvingly. Then a woman in a pretty white
blouse and a string of pearls pops up. Her dark hair is pulled up at
the sides, and she's wearing bright red lipstick. She's also cupping two
chicks in her hands – and then she lifts the chicks to her cheek and
smiles. 'Pretty chicks? Yes, sir!' the voiceover comments, enthusiastically, as the double entendre misses its mark. After that light relief,
it's back to the men for the 'enormous task' of wing-banding the
chicks with their flock number.

We follow the chickens through their short twelve-week lives, as
they grow into big, handsome birds, some brown, some with grey
stripes, some white as snow. They get put into crates for transport,
transferred to cages … and suddenly they're carcasses hanging from
hooks, ready to be judged. 'Twelve birds from each of the samples
were packed for display purposes. Others,' the voiceover explains,
'went on the eviscerating line.' Here are women again, pushing along
chickens strung up from their feet on what look like coat-hangers.
One man is in there, inspecting the birds. Then we see the final
display – a few exemplary cockerels in cages, and boxed-up chicken
carcasses draped with lametta. But outside, something extraordinary
is approaching: a chariot, covered in white fur, and flanked by two
American flags, carrying a woman in a white robe, wearing a crown.
Nancy McGee, described as a 'supplementary attraction to the
programme', is the Del Marva Chicken-of-Tomorrow Queen.

Still, Nancy doesn't detract attention from the real champions for
long. They're the small flock bred by Charles and Kenneth Vantress,
who crossed Red Cornish cocks with New Hampshire hens. It proved
to be a winning combination, securing the Vantress brothers first place
for both weight of birds, and efficiency of feed conversion into live
weight (translating into more food for your money). But this is a
beginning, not an end. The film isn't there just to announce results,
but to launch another national contest, to take place in 1951. More
men in suits are looking pleased about the prospect of future competitions. The voiceover wraps it up, 'Even today, housewives are enjoying

improved meat-type chickens' – and there are all the housewives, in a line, eating fried chicken thighs and drumsticks with their greasy fingers and grinning.

This film was clearly made in a very different world. A world where only men did serious work, while women either held fluffy chicks to their cheeks and looked decorative, or did the boring, mundane jobs. It's also a world where chickens were skinny things, while the poultry industry dreamed of making them what they are today: fast-growing, plump, white-fleshed monsters. The thing that hasn't changed is the approach. Right from its inception, the breeding of broiler chickens was clearly, already, an industry. How telling that chickens are referred to, in the opening sentence of the voiceover, as a 'crop'. And the genes of the winning chickens from 1948 are scattered amongst our commercial flocks today.

The victorious Red Cornish cross from the contest was bred together with a white-feathered Leghorn which had won in the purebred category. The result, the Arbor Acre breed, became immensely successful. What had been a small farm focusing on fruit and vegetables, with a sideline in chickens, became the major supplier of America's broiler companies. In 1964, Arbor Acres was bought by Nelson Rockefeller and exploded on to the global stage. Half of the chickens in China are descended from Arbor Acres lineages – descendants of the competition chickens. It sounds astonishing, and it's hard to imagine how breeding has changed chickens so quickly, and so completely.

The transformation of chicken production into an enormous global industry involved not only selective breeding on an unprecedented scale, but also extremely tight regulation of breeding. Today, chicken breeding and chicken rearing by farmers are completely separated. The very fact that chickens are laid in eggs which can be incubated by machines rather than hens allows for this complete division. Chicken farmers rear chickens – often on a huge scale – but they're not the ones breeding the chickens. That task is carried out by specific breeders – and just *two, huge* multinational companies dominate the market: Aviagen and Cobb-Vantress.

These companies keep a very tight rein on their pedigree breeding-bird populations. Three generations down the line from their protected pedigree flocks, they create 'parent stock' which is sold to broiler-breeder farms, where chickens from separate genetic lines are bred together to make the final mix. The resulting chicks are sent on to broiler 'Grow-Out' farms – even your free-range, organic birds can come from those industrial chicken breeders, though there are some smaller breeding firms that specialise in slow-growing chickens for the traditional and organic market. Most chickens, though, grow fast – and are slaughtered at just six weeks old. When we eat chickens, they're really just overblown, overgrown, big chicks. The ends of their bones haven't even begun to turn from cartilage to bone yet. A single great-grandmother hen, back in the pedigree flock, can have an astonishing 3 million broiler-chicken descendants – who never make it to adulthood.

As well as carefully controlling the characteristics of their pedigree chickens from a phenotypic point of view – scrutinising their growth trajectories, their weight, their feed consumption – chicken breeders are now using genomics to hone their selective-breeding techniques. But advances in genetics also hold out the possibility, not just of genotyping chickens, and identifying advantageous genetic variants, but of genetically *modifying* the birds. No commercial chickens have been genetically modified – yet. But the techniques are being tested out in research institutes. The tools to edit the DNA of chickens and other livestock – to remove deleterious pieces of DNA and to insert advantageous genes – already exist. It's been an arduous journey just to get to the point where the method works. Now the race is on to find ways of using that method to improve flocks. And just a seven-minute drive from the beautiful fifteenth-century Rosslyn Chapel in Scotland, mythologised in Dan Brown's *The Da Vinci Code*, is the Roslin Institute – where they're busy investigating a different kind of bloodline, a different kind of code. I travelled to Midlothian to meet these new code-breakers.

The researchers of Roslin

The Roslin Institute is a suite of state-of-the-art buildings, some designed to contain chickens, to get the best out of them, and others designed to hold scientists – to get the best out of them. The scientists here are focused on optimising their chickens – and not just through selective breeding. That's worked wonders on chickens over the last millennia, and then in a truly extraordinary way over the last sixty or so years. But now we can interact directly with the genetic code of an organism, selective breeding looks positively archaic in comparison. Domestication is a continuing process – and this is where the forefront of domestication currently lies.

New techniques in genetic modification promise the earth – literally. With their help, we could be farming in a much more efficient, sustainable, egalitarian way in the future. And yet we're afraid. Selective breeding is one thing, but for many direct genetic manipulation – using enzymes to modify DNA – seems a step too far, a Rubicon we should not dare to cross.

Instinctively, I feel there could be something wrong here. Science fiction has primed me – even me – to be wary of genetically altered organisms. Novelist and journalist Will Self is a master of writing about uncomfortable, unsettling otherness. In his *Book of Dave*, there are genetically modified pig-like animals called 'motos' which are both pets and livestock. They are intelligent; they talk – in toddler-ish broken phrases – but they will be slaughtered and eaten. The motos challenge our perception of the animals that we deliberately breed for consumption. We deem our tastebuds more important than their lives. There was too much dissonance there for me – I was completely vegetarian for eighteen years. Now I eat a little fish, managing my guilt, but other flesh is still a step too far.

We create a division in our minds between us and other animals – a necessary division if we are to eat them. You'd never consider eating another human (I imagine). But most people don't have a problem with farming animals, slaughtering them, and eating them. So what about *changing* them? This seems to be acceptable – if achieved through selective breeding. When it comes to plants, it seems we're comfortable with the idea of creating mutations using radiation or mutagenic chemicals, then selectively breeding these genetic changes into our crops. If that sounds novel and dangerous, it's actually something we've been doing regularly since the 1930s. More than 3,200 types of mutagenic plant have been created and released since then. Some of them are now grown and promoted as organic products. The majority of groundnuts grown in Argentina are bred from irradiated mutants. The majority of rice grown in Australia is bred on from an irradiated mutant type. Mutant rice is grown in China, India and Pakistan. Mutant barley and oats are widely grown in Europe. In the UK, Golden Promise barley, a mutant created by zapping plants with gamma rays, is grown to make beer and whisky. There's no danger at all from the radiation in the crops that are being grown – it's already done its work, scrambling DNA in their ancestors, and producing useful variants.

These plants are, quite clearly, all genetically modified. So why is it more acceptable to modify genes using an instrument as blunt as a beam of gamma radiation, whilst using an enzyme to do the same sort of thing – in a much more precise and controlled way – feels like it could be more dangerous? The International Atomic Energy Agency is keen to separate 'radiation breeding' from biological, genetic modification. Radiation breeding is described as being simply an accelerated version of the spontaneous mutation that occurs in organisms and which is the stuff of variation, the lifeblood of evolution itself. But if we're already modifying DNA using radiation, and calling that 'radiation breeding', it strikes me that we should be calling the – more exact and directed – biological version 'enzyme breeding'.

So I was keen to get inside the Roslin Institute and talk to the researchers themselves about their own take on genetic engineering, and the newest tools for doing it. They are pioneers, operating right at the frontier. They understand the science, as well as the swirling vortex of perception and prejudice and valid concerns, better than perhaps anyone else. And they know the genes of the chicken – the first domesticated animal to have its full genomes sequenced, back in 2004 – very well indeed. Adam Balic explained the techniques and their potential uses; Helen Sang talked to me generally about this science and the politics around it; and Mike McGrew told me about the exciting new developments – and his vision for this technology as a force for good in the world.

Adam met me and escorted me up to his light-filled office on the first floor of the steel, glass and copper-clad building that houses the scientists at Roslin. There were posters showing the stages of chick embryological development on the walls. We sat down and he pulled up images on the screens that took up most of the space on his desk. There were islands of bright green glowing against a black background. These were photographs, taken down a microscope, showing a developing chicken embryo. We were looking at its neck, and the patches of green were showing up a specific type of tissue – lymphoid tissue, the same sort of stuff that makes up our own lymph nodes. This tissue doesn't usually glow green: Adam had engineered the chicken embryo, inserting a 'reporter gene' into the chick's genome that would produce a green, fluorescent colour wherever lymphoid tissue developed.

He'd made this change to the embryonic chick's DNA using a traditional method, or at least, one that's been used for around twelve years in chickens. He'd used viruses to do the work for him. Many viruses work by inserting DNA into a host's genome, and so it's possible to hijack this mechanism, getting the virus to insert the gene that you're interested in, into the cell of another organism. These 'viral vectors' were originally developed for human gene therapy, but they work well for chickens, too. Although it's not generally possible to direct the virus to a specific position in the new genome, they seem to be pretty good at finding places to insert genes where that gene will have a good chance of being read, or expressed, by the cell.

Adam was using this tried and tested technique to illuminate lymphoid cells in his chick embryos. The way he'd done it was to identify a protein that was normally made in those cells, but not in others, and then to find the 'on-switch' – the regulatory sequence of code that sits just upstream of the code for the protein itself. Then he could construct a new length of DNA – combining that particular 'on-switch' with a gene for making a green fluorescent protein – originally isolated from a jellyfish. Using a viral vector, Adam could insert that whole new package – the switch plus the jellyfish gene – into the chicken embryo. Then, in any cell where the switch was thrown to make the normal lymphoid-cell protein, the gene for the glowing protein would also be switched on. The genetically modified embryo obligingly 'stained itself', revealing its lymphoid tissue with stunning clarity, when illuminated with UV light under the microscope.

'These aren't just pretty pictures. They will allow us to quantify things too,' Adam explained. The images showed precisely where lymphoid tissue – associated with the immune system – was developing in the embryo. Adam was studying the development of the chick immune system and these striking images were crucial to working out how the relevant immune cells and tissues formed. We were looking at how the chick's defences were being laid out, almost like mapping ancient fortifications and trying to understand how a battle was fought. Birds have a very different immune system from mammals, so unusual in comparison that it prompts us to question how they've managed to survive without the tools that mammals have developed?

'Almost everything we've learnt from mammals tells us that birds shouldn't exist,' said Adam, 'but they cope with the same environment, the same pathogens – coming up with different solutions.' Noticing differences like this, and trying to understand why those differences exist – that's often how science proceeds. Lymph nodes seem so critical to mammals, including us. Birds possess patches of lymphoid tissue, but nothing as discrete and definite as a lymph node – and they manage perfectly well without them. It's an interesting conundrum. Lymph nodes seem like quite complex things to invent. Why do mammals need them while birds don't? By default, we'll understand a lot more about the human immune system if we find out how birds manage to fight off infections with their quite distinct immune systems.

Genetic modification has enabled embryological development to be mapped more precisely than ever – it was clearly an important tool for fundamental scientific research like this. But what about the applications for genetic modification that could take it, out of the lab, into chickens bred for food? The Roslin researchers were looking at that angle too, using the combination of a quirk of embryological development together with an astonishingly precise new gene-editing technique.

Spreading a particular version of a gene through a flock of chickens relies on getting that gene into the cells which produce gametes – the eggs and sperm. The gamete-producing cells in the gonads of chickens (and humans) are known as primordial germ cells. They are essentially immortal cells – they will divide and divide, with some progeny 'growing up' into eggs or sperm, depending on the sex of the animal, and some staying as germ cells, ready to divide again – to make more eggs and sperm, and to replace themselves. The conventional way of getting your gene of choice into those primordial germ cells is indirect and more than a little haphazard – by selective breeding. You identify chickens with a particular trait, and breed those chickens together, hoping that the gene for the trait is there in some eggs and sperm, and will make it into some birds in the next generation. It takes generations to spread a trait through a flock. But imagine if you could short-circuit that process, by ensuring that all the eggs of a hen or the sperm of a cockerel contained the desired gene – then all their offspring will have that gene and exhibit that trait, straight away. This is precisely what the newest gene-editing tool allows the geneticists to do. And, serendipitously, it's relatively easy to remove primordial germ cells from a chick embryo, in order to modify them.

Chicks have fascinated embryologists ever since Aristotle followed the three-week development of hens' eggs. It's possible to raise a section of the eggshell and to observe the developing embryo – and even interact with it – without killing it. The embryo develops on one side of the ovum – and you'll be familiar with what that ovum looks like. Before it gets covered in albumen, and a shell, it's the yellow bit that is mostly a massive yolk.

While the ovulated hen's ovum measures about 2.5cm across, a human egg is just 0.14 millimetres in diameter. But that's actually still a very large cell, compared with the size of other cells in the body. It contains enough cytoplasm – the stuff inside the cell – to get embryonic development under way after fertilisation. The fertilised human egg is able to divide itself up into a ball of cells, without actually growing in size. In comparison, the hen's unfertilised egg is *huge*. It's the size of the yolk in a laid egg – and that's exactly what most of it is. One huge cell, stuffed full of yolk nutrients to support the developing embryo, with a tiny, tiny bit of cytoplasm at one end – it's there on your breakfast plate if you bother to look for it. In that cytoplasm are the chromosomes that represent the female genetic contribution to the embryo. The male genetic component is delivered to the egg by the sperm. And that's when things start to get interesting. Whereas mammal eggs divide slowly, with the first cell division, into just two cells, complete after approximately twenty-four hours, the fertilised chicken egg doesn't hang around. By the time the hen lays the egg – twenty-four hours after fertilisation – a disc of around 20,000 cells has already formed. If you opened up the egg immediately, you'd see it – a whitish disc on one side of the yellow egg yolk. If the laid, fertilised egg is kept nicely warm, the blastodisc (those 20,000 cells) continues to grow, multiply and develop into a chick embryo.

At just four days after being laid, the blastodisc has already rolled itself up into what will become the body of the chick. The developing eye is clearly visible and the chick embryo's heart is already beating. (The human embryo only reaches an equivalent stage of development a full *four weeks* after fertilisation.) A network of blood vessels has also developed around the chick embryo by this point, reaching out around the yolk of the egg. If you shine a light through a four-day-old, fertilised and incubated hen's egg, you can see these blood vessels quite clearly, radiating out like spidery red tendrils from a central red spot which is the embryo itself. If you were able to make an opening in the egg, and insert a tiny needle into one of those embryonic blood vessels at this stage, you'd be able to draw out a minute sample of blood. Within that sample there will be early blood cells – but also some extremely important stem cells. These are the primordial germ cells – which would eventually settle in the gonad of the developing chick, ready to make eggs or sperm, depending on the sex of the chicken.

Mike McGrew is working on taking blood from slightly younger embryos, only two and a half days old. At this stage, just a tiny sample contains 100 germ cells. The next trick he's pulled off is to get the cells to grow in culture, away from the embryo, for months and months. That gives him the opportunity to edit their genes – using a new technique to make precise modifications, cutting out some pieces of DNA and splicing new ones in.

Having made those adjustments, the primordial germ cells can then be injected into a chick embryo that has already been genetically manipulated so that it doesn't produce any of its own germ cells. Amazingly, development then proceeds as normal – the genetically altered primordial germ cells migrate to the ovary or testis of the developing chick. When that chick hatches and grows up into a hen or cockerel, that bird will be producing eggs or sperm which *all* contain the adjusted DNA.

The instrument which is allowing the geneticists to make precise adjustments to genomes is called CRISPR – the sharpest new tool in the genetic-engineering, neo-Neolithic toolbox. It's much more refined than the traditional, viral vector method, but it's also borrowed from nature, and based on years of painstaking research into the ways in which viruses and bacteria wage war on each other.

Some bacteria have a clever method of defending themselves against viral attack – a system which essentially provides them with immunity against viruses. When bacteria are exposed to viruses they copy a section of the viral genetic code into their own genome. It seems foolish – aiding and abetting the virus in this way – but it's not. It means they can 'remember' the pathogen and fight it off effectively the next time. The piece of pathogen DNA is flanked by strange, repeating sections of genetic code: bookmarks for the bacterium. It's these bookmarks that are known as CRISPR: Clustered Regularly Interspaced Short Palindromic Repeats. When the bacterial cell becomes infected, it looks up the bookmark and reads the short section of pathogen DNA – copying that sequence, using a slightly different molecule, RNA (which stands for ribonucleic acid, whereas DNA is deoxyribonucleic acid). That copy, the RNA 'guide', links up with a DNA-cutting enzyme in the bacterial cell which acts like a molecular pair of scissors. The RNA guide homes in and locks on to DNA arriving with an invading pathogen – and the enzyme neatly cuts it up, disabling it. So, if you wanted to make a very precise cut in a piece of DNA, you can specify your own target by creating an RNA guide – then giving it to the scissor-enzyme, to make a snip exactly where you wish. You can make as many cuts as you like, where you like.

The potential applications for this new tool are myriad. With this new gene-editing technology, it's possible to snip out particular genes much more precisely than ever before, creating a 'knockout' embryo. As this embryo develops it will reveal what the function of that gene would have been, by demonstrating what happens in its absence.

Understanding embryological development better will help us tackle diseases in the future – not only in chickens, but in vertebrates more generally, including in us humans. CRISPR could also be used therapeutically – to remove damaged DNA in living organisms. It's already been employed, in the lab, to remove cancer-causing pieces of viral DNA from human cells. In fact, the technique is so refined that it can be used to snip out a single base pair – effectively just one nucleotide 'letter' on a chromosome – from a genome. But it's not all about removing DNA – CRISPR makes it possible to precisely remove a section of DNA, and to splice in another. Cells are never happy about having their DNA snipped up. Molecular machinery will swing into action to repair the damage. The cell will usually look at the other chromosome in the pair to help it reconstitute the damaged DNA. But you can make a suggestion to the cell, by introducing your own piece of template DNA for it to copy instead. This has already been done – in the lab – to modify yeasts to make biofuels, to alter crop strains, and to engineer mosquitoes resistant to malaria. The American Association of the Advancement of Science branded this new gene-editing technique the scientific breakthrough of the year in 2015. The field is moving fast – the range of potential applications is dramatic – but ethical questions abound.

Helen Sang has been studying vertebrate development and using genetic-modification techniques for more than forty years. She's still interested in uncovering the fine detail of embryological development, but she's also worked on chickens to genetically modify them to produce valuable proteins – things they never normally manufacture. Helen had done this with hens' eggs and human interferon, a protein made naturally in the human body but also used as a drug to help fight viral infections. The egg white that hens make contains the protein ovalbumin. If you take the regulatory sequence, the 'on-switch' for ovalbumin, and join that up with the human interferon gene, you can stick that package into hens and they'll start to make interferon alongside ovalbumin. So it's possible to modify chickens to make it easier to study their development, as Adam had done with his green fluorescent protein in the lymphoid cells, and it's also possible to get chickens to make other useful proteins for you, in their eggs – such as that interferon.

But in recent years, the focus of Helen's research at Roslin had moved to looking at ways of modifying the chickens we eat. She was interested in something of immediate, real world relevance – promoting disease resistance in chickens. She was excited by the potential of CRISPR to achieve results precisely and quickly. She explained how it could work. You'd start by screening birds for disease resistance – to avian flu, for example – and then looking for genes associated with that resistance. That gene may only differ by a few nucleotides from the sequence in another bird, but those tiny differences can be crucial. Having identified a useful gene, you can use CRISPR to cut out the equivalent in another bird, and then replace it with the one you know will be beneficial. Using the technique in this way, you'd simply be spreading a genetic variant which *already existed* in chickens through a flock, without having to go through the laborious process of selective breeding. But of course there is another possibility: as well as introducing a variant of a gene from the same species, the same technique can be used to bring in a gene from a separate species. 'We can move genetic information from anywhere to anywhere,' Helen said, quietly in awe of the technique. 'I think that's a possibility which causes more worry – the idea that you move genetic information across the species boundary,' I observed. 'Well, it's all DNA,' replied Helen, 'and anyway we know that DNA gets around – you can find things in us which have come in from other species.' And that's true, you can – especially from viruses, which love to go sticking their genetic oar into other genomes.

In fact, it isn't just naturally occurring genes that geneticists could move from one species to another; they can now make entirely novel, artificial genes. It sounds extraordinary, but it's already bearing fruit in chickens, if that's not too mixed a metaphor. Geneticists are already exploring this avenue, designing artificial genes – specifically designed to throw a spanner in the works of viral replication – from scratch. 'If you know a lot about the flu virus, you can work out novel ways to derail it,' explained Helen. One promising gene has induced chicken cells to make a small RNA molecule which caused problems for viruses, but Helen's experiments have shown that it doesn't confer complete resistance – there's clearly still much more work to be done in the lab before a gene-edited, flu-resistant chicken is a reality. What

I was convinced of at Roslin is that it's certainly on its way, and not too far over the horizon.

Working on aspects of biology which have such obvious benefits as disease resistance may encourage acceptance of the use of genetic modification in livestock and crop development. Helen thought that CRISPR itself may help to allay some fears. The precision of the technique meant that you could insert a gene somewhere where it wouldn't derail any other operations in the cell – geneticists call such locations 'safe harbours' – while at the same time maximising its chances of being read, or expressed, by cells. With traditional modification, using viral vectors, you couldn't predict *where* the gene would be inserted – though of course you can check afterwards. But with CRISPR, you can go straight in and make sure that gene is placed exactly where you want it.

Helen was eloquent on public perceptions of GM. She believes that the debate has been hijacked by certain lobbying groups with rigid agendas, and that the public in general isn't being given the chance to choose: to accept the technology – or not. 'At the moment, GM isn't a choice. You can't go to a supermarket and buy a GM chicken. We don't sell anything that's genetically modified. It's very unusual to have a whole technology which is excluded, rather than giving people a choice.'

Helen told me that when she first started working in this area of science, and told people what she was doing, the reaction was generally positive – 'They thought it was really neat, a great idea. Then suddenly it became anathema as far as food goes.' I asked her if she thought that it all went back to the debacle that ensued when the biotech company Monsanto tried, controversially, to introduce its GM soy to Europe in the 1980s – and she did think that this played a significant role. Somehow, the debate over GM became hopelessly entangled with concerns about the domination of big, multinational companies. But this is something that worries Helen, too. 'There are many things that I would be just as concerned about as somebody from Friends of the Earth, about where our food is coming from,' she said, somewhat surprisingly, 'but I think it's a complete distraction to focus on GM. It's a technology which has something to contribute. And we should be able to find a way of allowing that to happen, and allowing people to make their own choices. GM's been used to epitomise bad, big business – whereas it's actually just another tool.'

Not only is the conflation of GM and big business unhelpful to working out how, as a society, we feel about this technology, Helen also believes that it's become a positive distraction from the real issues facing us about the future of food production. 'There's a smaller and smaller number of very large companies controlling food production. That's not a scientific problem – it's a political and economic one,' she explained. 'And it's a tricky one. You have to accept that it's been very efficient. We do need to feed a lot of people. But we need to have more sophisticated conversations about how we can take advantage of those efficiencies while protecting the environment and feeding the financial reward back to society.' In some ways, the concern over GM, and the ensuing, incredibly tight regulations placed on this technology, only makes that problem worse. The cost of meeting the regulators' demands is so high that it is effectively prohibitive. Only big multinationals can really afford to invest in GM – it's stifling innovation and placing it solely in the hands of a small number of huge companies.

I asked Helen a difficult question: what did she think might happen over the next decade? Was it possible that GM would become more acceptable? She thought so. Younger people seem less likely to reject the idea out of hand. 'But then, in the States, we're seeing a backlash,' she said. There have been moves in some US states to enforce labelling on GM foods – something they've never done previously. The idea of labelling something 'GM' is strange in many ways, especially if you're only going to include enzyme-induced modifications, and not the ones created by irradiating organisms. There's no risk to human health from eating GM food, even if you disagree with the methods used to produce it. And what does a 'GM' label tell you anyway? To be even moderately informative, it needs to describe what the modification is, and what its consequences are. 'But on the other hand, if people want to know, then they have the right to know,' said Helen. 'It's a really tricky argument.'

We discussed Golden Rice – a GM form of rice with enhanced levels of vitamin A, designed to combat dietary deficiency – which has provoked such a variable response from the public. Some accept it as a genuine philanthropic effort, and believe that it could really help to reduce vitamin A deficiency, particularly in some of the poorer countries of the world. Others see it as merely a poster child, something the 'GM industry' has bought into, simply as a tool of persuasion – the acceptable face of GM and the thin end of the wedge. It seems entirely reasonable to distrust the motives of some large companies who are trying to sell more of their own herbicide off the back of selling GM crops. But perhaps we should be more trusting of efforts to help the poorest farmers and communities – as Bt Brinjal, a disease-resistant GM aubergine produced as an entirely not-for-profit exercise, has already done. 'If we want to be producing food efficiently and sustainably, we shouldn't cut ourselves off from some of the ways we can do that,' Helen concluded.

Perhaps it's too easy to be cynical. It seems a shame that this technology wasn't first developed and implemented by universities or not-for-profit enterprises. I have no doubt that there wouldn't have been such a backlash and a collapse of trust if that alternative historical scenario had played out. But genetic modification has become so tainted by its connection with big business and questionable motives. It's hard to shake that off – even where research is now being carried out by publicly funded, university research institutes.

For Mike McGrew at the Roslin Institute, one of the most exciting prospects for gene editing, if it is ever allowed out of the lab into the real world, lies in its potential for promoting disease resistance in farmed animals, particularly in developing countries. 'We're working with the Bill Gates Foundation in Africa,' Mike told me, with undisguised pride. 'Anything that would make these commercial chickens survive and thrive better over there, and actually lay eggs in a non-ideal climate, would be such a huge benefit.' But Mike's not just interested in the potential of these new techniques for breeding better commercial poultry, particularly in Africa, but in possible applications amongst wild birds too.

'This is the thing I really care about: conservation biology. Think about the honeycreeper that lives on the Hawaiian Islands. We effectively brought in avian malaria to Hawaii, and the native honeycreeper has no resistance, because it's never seen avian malaria before.' All the birds living at lower altitudes have been killed off. The only ones left are those high up in the mountains, where it's cooler and the mosquitoes don't survive. But now, with global warming and rising temperatures, the mosquitoes are starting to reach higher altitudes, placing honeycreepers under increased threat of extinction. 'So imagine if we were smart and we knew the genes responsible for resistance to avian malaria,' Mike mused. 'Could we could go into these wild populations, edit their genes, then release them? So there would now be disease resistance and the honeycreepers would thrive. Just imagine doing that.'

Mike understands the antipathy to GM when it's focused on food in developed countries. 'But if you're able to do something useful for the human race, for the planet – there are a lot of different things we can use this technology for – I think people will start to recognise and welcome the potential.' He spoke with real passion, but without any hype. 'We need more education,' he said. 'Not fake news on the internet or in tabloids. People think DNA is the essence or the soul of an animal – and we're changing its soul. But when they understand what DNA actually is, and what this technology is, they stop being scared.' And yet it seems unlikely that the first GM chickens will hatch out of the broiler industry. The commercial firms are too sensitive to lobbyists. And with the Food and Drug Administration in the United States now keen to label any genetic modification – even to a single base pair – as requiring the same level of regulation as a new drug, the technology is unlikely to take off in America. So where did Mike think the first genetically modified chickens would finally enter the human food chain? 'It'll be China,' he said. 'China without a doubt. They have the genetics. And they have bird flu.' If I was the betting type, I'd also be putting money on that immediately. I'm sure Mike will be proven right. If not in 2018, then very, very soon.

Which came first?

We can speculate about where the first commercial, directly genetically modified chickens will emerge. But we've been indirectly editing the chicken genome for centuries – where and when did that all start? The answer provides us with the solution to an enduring question that has left our best philosophers completely and utterly stumped. A riddle that addles the mind and draws one into a downward spiral with insanity seemingly the only end point.

Which came first? The chicken or the egg.

And here's the thing. Evolutionary biologists have an answer for this question. Because, before the chicken, there was the junglefowl – and that laid eggs too. And so did its ancestors, all the way back to the dinosaurs, and still further back in time. Eggs, clearly, came first.

With that epic question answered, we still need to pin down the actual origin of chickens. In the 1990s, researchers seemed pretty sure that all chickens came from a single origin, that the ancestral species was the red junglefowl (as Darwin had, once again, rightly predicted), and that domestication had occurred in a discrete area of south or south-east Asia. The genetic diversity of modern chickens is highest across that region, and much lower in China, Europe and Africa. Some researchers have suggested that the chicken homeland was, very specifically, the Indus Valley, 4,000 to 4,500 years ago (2000 to 2500 BCE), during the Bronze Age. References to 'the bird of Meluhha' in cuneiform tablets from Mesopotamia, dating to 2000 BCE, could relate to chickens – Meluhha itself is thought to be an ancient name for the Indus Valley. Others, though, have favoured an origin further east. Today, several distinct subspecies of red junglefowl scratch around in forests stretching throughout south and south-east Asia, from India, Sri Lanka and Bangladesh, to Thailand, Myanmar, Vietnam, Indonesia and southern China.

Doesn't this sound like a familiar story? We know what comes next. As more information came in from wider genetic studies, the theory of chicken origins was rewritten – as one which involved diverse geographic centres of origin, across south and south-east Asia. But that impression of multiple origins is also compatible with a single origin – perhaps across a relatively wide area – followed by dispersal and extensive interbreeding with wild species along the way. Modern chicken genomes contain strands of ancestry woven in through interbreeding with closely related birds including other subspecies of red junglefowl, as well as different species such as the grey junglefowl and Ceylon junglefowl.

In 2014, a piece of research published by Chinese geneticists really put the cat among the pigeons, or the fox among the chickens, for want of a better metaphor. They articulated an astonishing claim – that chickens had been domesticated by 8,000 years ago on the North China Plain. It was so out of kilter with the rest of chicken science – and feathers flew. Most researchers remained sceptical, however, for several reasons. Firstly, the climate of the North China Plain 10,000 years ago was decidedly unsuitable for tropical junglefowl – the accepted wild progenitors of chickens. Secondly, the identification of bones from archaeological sites on this plain seemed rather suspect. Some bones thought to be those of chickens were probably pheasants. And other bones appeared to have been entirely misidentified – not belonging to any bird, even, but to dogs. It seems that this was an extraordinary claim – which turned out to lack the requisite robust, extraordinary evidence. South and south-east Asia remained the most likely homelands of chicken-kind. And from that starting point, chickens set off to conquer the world.

Pacific chickens

Thousands of miles away from the homeland of chickens, these birds have been drawn into an ongoing debate about the human colonisation of the Americas. The premise is this: if chicken history is so tightly bound up with human history, then elucidating events in the deep past of chickens will shed light on what humans were up to as well. Reconstructing the population movements that led to the colonisation of the Pacific has been a tricky challenge. The peopling of Pacific Islands happened relatively recently, within just the last 3,500 years – but waves upon waves of colonisers have left confused trails behind them. The challenge is like looking for footprints in the sand, to recover the evidence of these ancient journeys. Imagine standing on a popular beach in Britain at the end of a summer's day, just as the last families are packing up their windbreaks, their towels, and their buckets and spades, and heading off. If you mapped all the footprints on the beach, could you reconstruct all the events of the day? Could you work out how many people had been there, from which direction they'd walked on to the beach, and roughly what time they'd arrived? It would be a huge challenge.

Reconstructing ancient migrations is an even more daunting task. And yet, with archaeological and genetic evidence combined, it is just about possible. Humans didn't arrive on the remote islands of Oceania alone; they took various other species with them – some deliberately, others less so, but all of them have a story to tell. Geneticists have attempted to track the human expansion into the Pacific by studying the molecular secrets hidden away in species as diverse as bottle gourds, sweet potatoes, pigs, dogs, rats – and chickens.

The islands of Near Oceania, in the south-western Pacific, were colonised way back in the Pleistocene, over 30,000 years ago. But Remote Oceania – including the groups of islands also known as Micronesia and Polynesia – wasn't peopled until much later, in the Neolithic. It was the last major migration of humans to entirely unpopulated lands. Archaeologists and linguists have suggested that this colonisation had happened in two waves, with an earlier migration of farmers, bearing characteristic Lapita pottery, starting around 3,500 years ago, and a later one, around 2,000 years ago. But chickens don't seem to bear witness to this two-wave model. Studying mitochondrial DNA from both modern and ancient chickens, geneticists found a distinct signature stemming from a single prehistoric introduction of chickens into Polynesia. The picture was extraordinarily clear – there was a founding lineage, from which all later Pacific Island lineages had evolved. The mitochondrial lineages of chickens, from the Solomon and Santa Cruz Islands in the west to Vanuatu and the Marquesas Islands further east, hark back to the original prehistoric arrival of farmers – and their fowl – in the Pacific Islands. For a while, genetic studies of humans also suggested that the colonisation had happened in one wave, but recent analysis of ancient human genomes has lent new support to that two-wave model suggested by the spread of material culture and languages amongst the islands of Polynesia. The chickens, it seems, were leading us up the garden path – and not for the first time.

For a while, it seemed as though that eastwards spread of farmers – and their chickens – might even have kept going right across the Pacific. The identification of a particular mitochondrial DNA type in chickens both from Rapa Nui (Easter Island) and from South America suggested just such a link. This was exciting, and controversial, as it indicated pre-Columbian contact between Pacific Islanders and the Americas. But the latest research on chickens, with exquisitely careful checks to rule out contamination, have revealed no such link. The DNA of Rapa Nui chickens and South American chickens is, in fact, quite distinct. South American chickens are essentially an offshoot of European chickens, which fits with the – much less controversial – idea of a post-Columbian introduction from Europe. This isn't to say that there was no early contact between Pacific Islanders and South America, though – sweet potatoes made their way from South America to Polynesian islands long before the arrival of Europeans in the New World. And the genomes of modern-day Easter Islanders show traces of admixture with Native Americans, dating to between 1280 and 1495 – whereas Europeans only reached the island in 1722. But this is still only circumstantial evidence: what's really needed to clinch this one is DNA evidence of admixture from pre-Columbian bones, either from the Americas, or from Polynesia. And that, for now, is proving elusive.

Genetics adds an important line of evidence, enriching and complementing the story that we can obtain from archaeological, linguistic and historical data – but not displacing those sources. Each provides us with a separate perspective on ancient reality. But when we contemplate prehistory in this way – staring at the past through such a wide lens – it's easy to forget that these were people, and animals and plants, just as we encounter them today. Not species, but individuals. Science is powerful – it can answer our questions – but sometimes I feel the chill of the abstract quite acutely. We build knowledge in this way, certainly – but perhaps, at times, we lose sight of the personal, the intimate, the moment.

Considering the people, then – we can imagine early farmers making voyages into the Pacific and settling on islands alongside hunter-gatherers. And there would undoubtedly have been a two-way flow of information. The hunter-gatherers shared their local knowledge of plants and animals – where to find them, and what was good to eat. The farmers shared their knowledge too – and their domesticates. It may not always have been this friendly, but by and by, the hunter-gatherers adopted more of the farmers' ways, and started to grow crops and breed animals. Gradually, and probably without any definite decision to do so, they became part of the Neolithic Revolution.

Early birds in the west

Chickens were not part of the original spread of people and ideas and livestock that marked the beginning of the Neolithic in Europe – they were domesticated too late for that. By the time chickens made their way into Europe, the Bronze Age had dawned. By 2000 BCE, chickens had spread from the Indus Valley into Iran. From the Middle East, chickens could have spread via a coastal route, to Greece and across the Aegean into Italy. Maritime trade had really taken off by the Bronze Age – this was the era of the Mycenaeans, the Minoans and the Phoenicians. The Mediterranean was crammed with merchant ships plying their trade. An alternative route may have seen chickens spreading north from the Middle East, through Scythia, and then west into central Europe. But it's also possible that some chickens could have spread from much further east – from China, and then via a northern route, through southern Russia, into Europe.

Several researchers have suggested that differences between the chickens of northern and southern Europe reflect these two distinct routes of introduction. But once again, the history of this domestic species is horrendously complicated – and knottily tangled up with human history. It's hard to trace those first migrations of chickens into Europe. Since those first feathery pioneers arrived, chickens have been subject to natural and artifical selection; flocks have been lost to diseases and replaced; birds have been brought in from further afield. The chicken breeders of the late nineteenth century picked and chose, breeding for particular traits, creating hybrids and generally mixing up the genetic history of European chickens, until they got what they wanted. And yet it is possible to disentangle the threads – the history is still there, embedded in the DNA of living birds.

A huge survey of chickens from the Netherlands – including sixteen 'fancy breeds' as well as commercial varieties – produced fascinating results. Most of these chickens had mitochondrial DNA which formed a neat cluster with that of chickens from the Middle East and India. The Indian subcontinent is the likely geographic origin for this cluster of maternal lineages. But a handful of breeds had mitochondrial DNA which was characteristic of chickens from the Far East – from China and Japan. These included three Dutch fancy breeds – the Lakenvelder, Booted Bantam and Breda Fowl – as well as some commercial egg-laying breeds from the US. It's tempting to imagine that these breeds, with their Far Eastern mitochondrial genes, provide some support for the idea of a pioneering northern route into Europe. But in fact this handful of eastern lineages, which are not even that closely related, is much more likely to reflect more recent genealogy. It's likely that these erratic traces of East Asian ancestry have a very short history, deriving not from that first wave of chickens arriving into Europe during the Bronze Age – but from exotic birds imported much later, by nineteenth-century breeders. So far, then, genetic studies of chickens fail to provide any support for that northerly route from the Far East. Instead, the main flow of these familiar fowls into Europe was via the Mediterranean.

The first evidence of chickens in Britain dates to the late first millennium BCE, in the Iron Age, but it was the Romans who really made chickens popular in this corner of north-west Europe. Chickens are by far the most well represented of any bird species in Romano-British archaeological sites. And yet the evidence is still rather thin on the ground, especially compared with mammal bones – like those of pigs, sheep and cattle. Bird bones are relatively fragile, and easily crunched to bits by scavengers, so it's somewhat surprising to find any surviving at all. In rural settlements, away from centres of power and Roman influence, there's not much evidence of chickens. But we glimpse them in more Romanised sites – in towns, villas and forts. Given the slim chances of chicken bones surviving, this evidence suggests that chickens – and their eggs, of course – may have been an important food, at least for the elite, in Roman Britain. It seems that chickens became equally popular further north, beyond the reach of the

Romans. In South Uist, in the Outer Hebrides, there's a little evidence of chicken bones dating back to the Iron Age, though it's not until the subsequent Norse period that there's more widespread evidence of domesticated chickens, braving the chill of the Hebridean Islands.

Although it's tempting to assume that evidence of domestic chickens is evidence of people eating them, and their eggs, we shouldn't rush to conclusions. It's been suggested instead that the initial spread of domestic chickens across the Middle East, and then into Europe, was in fact less about meat and eggs – and more about blood sport. Images of cockerels fighting appear on seals and pottery from Egypt, Palestine and Israel, dating to the seventh century BCE. The sport was popular in Ancient Greece, and also seems to have been exported across the Roman Empire. Archaeological collections of chicken bones from Velsen in the Netherlands, and from York, Dorchester and Silchester in Britain, contain surprisingly high proportions of cockerels. Artificial cock-spurs have been found at Silchester and Baldock, but it seems that Ancient Britons may have been indulging in cockfighting even before the Romans arrived. Julius Caesar, in his *Gallic Wars*, wrote that the Britons 'regard it as unlawful to eat the ... cock ... but they breed them for amusement and pleasure'.

The idea that the spread of chickens across Europe may have been about something other than meat is supported by a couple of lines of evidence. Firstly, during the Middle Ages, chickens were relatively small – suggesting that meat may not have been the foremost concern of breeders. Perhaps keeping hens for eggs and cocks for fighting was more important. And there's written evidence too: geese and pheasants were much more frequently to be found on medieval menus than their now more popular cousins.

Domesticated chickens were transformed during the twentieth century, through the systematic approach to selective breeding largely unleashed by the Chicken-of-Tomorrow competition. But even before this, chickens had started to plump up, diverging away from their red-junglefowl ancestors. In just the last few years, geneticists have been able to identify particular regions of the genome that appear to have changed over time, and which seem to be linked to a size increase. They're also able to estimate *when* these changes to the genome took place. A study of modern chickens from around the world revealed that they all carried two copies of a particular variant of a gene associated with metabolism. The gene in question made a protein which was a receptor for thyroid stimulating hormone (TSH). The particular version of the gene which had become ubiquitous in modern chickens made them nice and plump. It seemed like a gene variant that surely must have been associated with the initial domestication of chickens – as essential as large seeds in domesticated wheat or corn. And yet the DNA of chickens from before a thousand years ago was almost completely devoid of this variant. It was only during the Middle Ages that the gene suddenly became much more common, sweeping through chicken populations.

This sudden spread of a gene for plumpness coincides with an equally sudden and significant increase of chicken bones at European archaeological sites in the tenth century, from 5 per cent to almost 15 per cent of animal bones. This seems to tie in with a religious and cultural change – the Benedictine Reform – which prohibited the consumption of four-legged animals during fasts (which could take up a third of the year) but permitted two-legged creatures, as well as eggs and fish, to be eaten. Suddenly fatter chickens became incredibly desirable, and human-mediated natural selection then worked its wonders, promoting the spread of that metabolic gene variant through chicken populations. Urbanisation probably played a role as well: although city-dwellers would have relied heavily on rural agriculture produce, they could also keep some animals – such as goats, pigs and chickens – in their own backyards.

Hormones can also affect the way animals behave, as well as their metabolism, and are implicated in an extremely important element of domesticated-chicken behaviour: a complete failure of maternal instinct. This sounds as though it should be bad for survival – and in the wild, it certainly would be. A hen who walked off from her eggs after laying them wouldn't stand much of a chance of passing her genes on to the next generation, but in domesticated hens, that's exactly what we want them to do. A hen that goes broody, that sits on her eggs and stops laying, is never going to win any prizes for egg production. The original red junglefowl lays fewer than ten eggs a year, whereas the most productive of our modern, domesticated egg-layers manages 300. That's only feasible because the instinct to incubate the eggs has somehow been bred out of our chickens. That possibility only arose when chicken farmers discovered the trick of artificial incubation. The earliest egg incubators are very ancient – going right back to Ancient Egypt. But the genetic changes associated with the most profound loss of maternal behaviour in chickens appear to have happened much more recently. The loss of broodiness in chickens is the equivalent of the non-shattering rachis in wheat and maize – incompatible with successful reproduction in the wild, but advantageous under domestication.

Geneticists set out to identify the genetic basis of this change in behaviour. They compared the genomes of two breeds of chicken with widely differing levels of maternal instinct: White Leghorns, well known for being prodigious egg-layers and lacking incubation behaviour, and Silkies, who like to sit on their eggs. They found two particular regions in the genome which were substantially different between the two breeds, one on chromosome 5 and another on chromosome 8. Both of these areas were involved – once again – with the thyroid hormone system; and the one on chromosome 5 contains the TSH receptor gene itself. Some changes to this gene were spreading through chicken flocks a thousand years ago – and are now found both in chickens bred for egg-laying, and in broilers, bred for the pot (or more commonly now, the oven). But other changes to the TSH receptor gene may have arisen more recently, explaining differences in egg production and maternal behaviour in modern breeds like the White Leghorn and the Silkie. It seems that tinkering with the thyroid hormone system in chickens may have killed two birds with one stone; or rather, caused two phenotypic changes with one genetic stroke. Once again, we're seeing how selection for one particular trait may influence another – this single gene seems to affect both plumpness and egg-laying in chickens.

These relatively recent changes to genes, bodies and behaviour remind us that domestication is not in fact a single, discrete event, but an ongoing process. And the arrival of gene editing means that useful changes can now be brought about even more swiftly than could be achieved in the tenth century, by papal decree.

7

RICE

Oryza sativa

He heaves his hoe in the rice-field, under the noonday sun,
Onto the soil of the rice-field, his streaming sweat beads run.
Do you or don't you know it? That bowl of rice we eat:
Each seed, each grain, the fruit of his labour done.

Li Shen, 'Pity the Peasants'

Feeding the world

Travel to Longsheng in the Guangxi Province of south-west China today, and you see a landscape transformed by agriculture, where people are still practising a centuries-old way of life. Steep hills rise up from a snaking river valley – and every slope is grooved with terraces. The tortuous, stepped paddy fields create the impression of something alive: a great, sleeping serpent. The Longsheng mountain range practically writhes, and the terraces are like the scales on its flanks. Longsheng means Dragon's Back.

I visited the rice terraces some years ago, and met a farmer, Liao Jongpu, whose family had cultivated rice there for generations. It was early summer, and we went up on the hillside with baskets of rice seedlings, to plant out the new crop on freshly ploughed terraces. Below us, more terraces were being prepared. Ploughing these narrow, winding strips defied an industrial, mechanised solution, but a single ox, pulling a plough, could navigate them easily.

Liao showed me what to do – taking three or four seedlings at a time, and pushing them down into the wet, yielding mud under the water. The seedlings looked just like grass – and of course that's what they *are*. Like wheat, rice belongs to the grass family, the Poaceae, and it looks similarly unpromising as a food – yet it's become one of the most important cereals feeding our huge global population. Rice contributes around a fifth of the calories and around an eighth of the total protein consumed worldwide. Some 740 million tons of rice are produced each year, and it's grown on every continent except Antarctica, and although it's also becoming an increasingly important staple in both sub-Saharan Africa and Latin America, around 90 per cent of the world's rice is grown and eaten in Asia. More than 3.5 billion people across the globe depend on rice as a staple, and it's the most important food crop in low- and lower-middle-income countries. For the poorest 20 per cent of the tropical population around the world, rice provides more protein per person than beans, meat or milk.

Across many low-income countries, the spectre of malnutrition stalks populations. Around the world, a billion people are starving and a further 2 billion suffer from 'hidden hunger', lacking essential micronutrients, including vitamins and minerals. The three most prevalent micronutrient disorders involve iodine, iron and retinol – or vitamin A.

Vitamin A deficiency leads to an increased susceptibility to infections. Malnutrition and infectious diseases often coexist – and each exacerbates the other. A vicious circle starts to spin as malnourished bodies fall prey to infection; infectious diseases suppress appetite and affect absorption of nutrients from the gut; the body's defences fall away. As well as this sinister synergy with infection, a lack of vitamin A is also one of the most important causes of preventable childhood blindness, responsible for around half a million cases every year. Half of these children die within a year of losing their sight. Vitamin A is found in animal products such as meat, milk and eggs. Where these foods are rarely eaten, vitamin A deficiency is likely to be more prevalent. Beta-carotene, a precursor of the vitamin, is found in some plant foods, including green vegetables, and orange fruits and vegetables – but conversion into vitamin A in the human body is fairly inefficient. So you need to eat a lot of those types of plant food to get enough of the vitamin, and for many people in poor countries that's just not an option.

Public-health strategies for reducing the burden of vitamin A deficiency have included schemes to encourage people to change their dietary habits, to grow their own carotenoid-rich foods – like leafy vegetables, mangoes and papaya – and to provide vitamin A supplements to children and breastfeeding mothers. Another way of boosting vitamin A intake is to fortify foods which are widely consumed, but naturally low in the vitamin. Breakfast cereals and margarine are often fortified with vitamin A in high-income countries. But in lower-income countries, this isn't a viable strategy as the poorest people are unlikely to have access to such processed foods.

There is yet another way of injecting more vitamin A into a dietary staple, not by processing it, but by inducing plants to add vitamin A, or at least, a precursor, to themselves. Genetic modification provides the opportunity to do just that – and rice, as such a globally important crop, presented itself as the perfect vehicle.

After eight years of work, the creation of a genetically modified form of rice that could produce its own beta-carotene was announced in the journal *Science* in 2000. Field trials started four years later in the US, followed by others in the Philippines and Bangladesh. Meanwhile, studies looking at the effects of consuming the rice concluded that it was safe to eat, and that just a small cup could provide half of the daily requirement of the vitamin A precursor.

But right from the start, Golden Rice attracted controversy. Greenpeace led the opposition, with concerns about Golden Rice being used as a public-relations exercise for genetic engineering – a superficially humanitarian initiative which would open the door for more profitable GM organisms. They said that Golden Rice was 'simply the wrong approach, and a risky distraction away from real solutions', whilst posing unpredictable environmental and food-safety risks.

In 2005, the project manager for Golden Rice, Jorge E. Mayer, met the criticisms from Greenpeace with a robust rebuttal. He was dismayed that a new version of Golden Rice, producing more than twenty-three times the level of beta-carotene contained in the prototype, continued to elicit opposition and disdain from environmentalists. He accused Greenpeace of disregarding evidence and sticking to an 'anti-biotechnology' agenda. To Mayer, Greenpeace and their allies were clearly the new Luddites, fighting back against a new, agro-industrial revolution. He wrote:

Nobody has been able to come up with a scenario whereby the provitamin A-enriched grains of Golden Rice could pose a menace to the environment or to human health. What's left in the opponents' camp is a perceived risk of the technology as such, rooted in unfathomable, yet-to-be-articulated dangers. Meanwhile, real threat does exist: it is the threat of widespread micronutrient deficiencies killing millions of children and adults all over the world.

Critics of Golden Rice had suggested that existing fortification and supplementation programmes could be derailed by a less successful venture. Mayer argued that this criticism failed to acknowledge the potential of GM rice to be a sustainable, and cost-effective, solution to vitamin A deficiency. Neither did it recognise the failure of existing programmes to reach remote, rural areas – where the need was most pressing. Mayer raised his own ethical objections in return – how could it be morally defensible to oppose something which could clearly have such a positive impact on human health, in some of the poorest communities on earth? He questioned how governments – and in particular, the European Union – could act on what he saw as such fragile evidence, meeting this potentially beneficial advance with regulatory knots so restricting, they threatened to strangle it.

Golden Rice has become something of a poster child for the pro-poor credentials and the potential of GM, and the biotech industry is also increasingly keen to cast itself as eco-friendly. But there's a suspicion in some camps that while GM developers are trying to present themselves as sustainable, progressive and caring – actually, this is all about a mob of fundamentally selfish corporations lining their own pockets. Trust had been broken. And the battle lines had been drawn up decades before Golden Rice arrived on the scene.

The creation of a monster

The largest corporation in the business, Monsanto, has presented confusingly different messages about the role of GM crops in global agriculture. In 1990, Monsanto's chief scientist, Howard Schneiderman, wrote about GM technology, stressing its many advantages, while cautioning that it would not be a panacea, a one-size-fits-all solution to the world's agricultural needs – and that it shouldn't be used to push farmers towards monoculture and cash crops. But at the same time, the Monsanto corporate giant was ploughing its own furrow, deeply and purposefully. It focused intently on a few, standardised weedkiller- and insect-resistant varieties of cotton and maize – that were *explicitly* designed to function as monocultured cash crops.

Anthropologist Dominic Glover has traced this disconnect between a pioneering scientist's vision and corporate practice right back to the emergence of Monsanto as a biotech giant. In the 1970s, Monsanto was in petrochemicals – including ones used in agriculture. This had become a risky business. Profits were tightly linked to the price of oil, with slim margins at the best of times. The Green Revolution had taken agriculture to another level, with new varieties of cereals, novel irrigation systems, pesticides and synthetic fertilisers supporting a doubling of production between 1961 and 1985. But after decades of innovation, it was becoming increasingly difficult to find new agrochemicals which would be more effective than the previous batch.

Monsanto had also run into difficulties as some of the chemicals which it had produced – including dioxins and PCBs – had proven to be damaging to both human health and the environment. The lawsuits started to pile up, and the future of the company was in jeopardy. Monsanto's survival depended increasingly on just one herbicide – their global bestseller, glyphosate, or Roundup. It had been a huge commercial success, but Monsanto couldn't rest on its laurels – the patent for glyphosate would eventually expire. The company needed to broaden its horizons.

In 1973, Stanley Cohen and Herbert Boyer had created the first transgenic organism – where DNA is moved between species – by taking a section of genetic code from one bacterium, and introducing it into another. Biotechnology – and specifically genetic modification – was looking like a promising area to explore, and invest in. Monsanto shed its chemicals and plastics divisions and reinvented itself as a biotech pioneer. And its first foray into commercial GM crops emerged in the form of a glyphosate-resistant, 'Roundup Ready' soybean – thus shoring up the market for Roundup as well. If you sowed the GM soy and doused your fields with Roundup, all the weeds would die – but the soy would thrive. In 1994, Roundup Ready soy was approved for agricultural use in the US. In 1996, Monsanto attempted to launch the soybean in Europe. And they couldn't have chosen a worse moment.

Suspicions about industrial agricultural practices and government were running fearfully high. Ten years earlier, an epidemic of BSE (bovine spongiform encephalopathy), also known as mad cow disease, had broken out in British cattle. This terrifying, untreatable disease would – after a long incubation of several years – cause a cow to trip and stumble, to become aggressive, and finally to die. The epidemic lasted from 1986 until 1998.

The origin of the disease was eventually traced to calves being fed with protein supplements – meat and bone meal – which were themselves contaminated with the remains of scrapie-infected sheep. Feeding meat and bone meal to cattle was banned, and millions of cattle were culled. But potentially hundreds of thousands of infected cattle had already passed into the human food chain. There was a fear that humans could be affected, through eating contaminated beef, but the British government sought to reassure the worried populace. In 1990, the agriculture minister, John Gummer, went as far as to publicly eat a hamburger, with his four-year-old daughter Cordelia, in order to demonstrate the safety of British beef. But then people started to suffer from something that looked suspiciously like a human form of BSE, causing stumbling and tremors, and ultimately leading to coma and death. The brains of sufferers became porous and spongy – just like those of cows with BSE. Further research proved the link between BSE and the human version, known as variant Creutzfeldt-Jakob disease, or vCJD. Although the number of deaths from vCJD was very small compared with other threats, peaking at twenty-eight deaths in a year in 2000, there was something peculiarly harrowing about the way it struck down its victims.

The British government finally acknowledged that BSE-contaminated beef could represent a risk to human health in 1996, but public trust in industrial agricultural practices, and in government, was in tatters. Enter Monsanto. EU officials approved imports of its GM soybeans in 1996, but British consumers were deeply suspicious. Tabloid newspapers picked up the scent. In 1998, Prince Charles wrote an article, in the *Telegraph*, titled 'Seeds of Disaster', in which he warned that transferring a gene between one species and another 'takes mankind into realms that belong to God, and to God alone'. Greenpeace launched itself into high-profile campaigns. The fear of genetically modified organisms, or GMOs, started to ripple out. They were seen as monstrous creations, signs of science run amok, and labelled 'Frankenfoods' by the media. Supermarkets across Europe banned foods with GM ingredients.

Monsanto reacted with its own advertising campaign, pushing the humanitarian potential of GM hard, asserting that 'worrying about starving future generations won't feed the world. Food biotechnology will.' In 1999, Monsanto CEO Bob Shapiro spoke at the fourth annual Greenpeace Business Conference. Shapiro said he wanted dialogue, not debate. He opined that Monsanto were indeed guilty – of believing too passionately in the beneficial impact their technology could have. It sounded too much like a faux apology. He stressed the potential benefits of biotech – in reducing water use, soil erosion and carbon emissions – but to many these sounded liked empty promises when the GMO that Monsanto was keen to introduce to Europe was a weedkiller-resistant soybean. However productive that soy turned out to be, it just looked like a way of Monsanto selling even more of its own, bestselling weedkiller. Even Robb Fraley, one of Monsanto's top in-house scientists, is said to have lamented, 'If all we can do is sell more damned herbicide, we shouldn't be in this business.' The gap between rhetoric and practice had never seemed so stark.

At the same conference, Peter Melchett, the executive director of Greenpeace UK, declared that '[the] public have taken a clear look at what you are offering and have said "no". People are increasingly mistrustful of big science and big business.' He went on to predict not only a European, but a global rejection of GM, based on civilised values and respect for the natural world. And he was right. The opposition quickly grew into a worldwide phenomenon. In 1999, Deutsche Bank analysts announced that 'GMOs are dead'.

In 2006, following actions brought by the US, Canada and Argentina, the World Trade Organization ruled that the European Communities had acted unlawfully by imposing a de facto moratorium on GM food, and that concerns about risks to public health were not supported by the scientific evidence. But governments weren't the only ones putting up barriers to trade: consumers and supermarkets continued to resist. BSE had sensitised European consumers to risks – especially any which were linked to big business.

Monsanto's image, always a little grubby, was transformed into something quite diabolical. Search #monsantoevil on the internet and you'll get a glimpse of the hatred and distrust directed at this techno-logical giant. And inextricably bound up with this 'supervillain' image is the technology itself – genetic modification – and those 'seeds of disaster' that mere humans should never be arrogant enough to sow. It seems that an inauspicious launch of a herbicide-resistant soybean, together with entrenched distrust of big science and big business, may have successfully hobbled this technology.

There's something poignant and ironic in that speech of Shapiro's at the Greenpeace conference at the tail-end of the last millennium. The Monsanto chief said that the way forward was dialogue, not polarised debate. Perhaps if the corporation had started in that way, right at the very beginning of their biotech research programme – engaging in real two-way communication and partnership with farmers and consumers – then the story would have been very different. Instead, they were so sure of the benefits of GM – with their chief scientist calling it 'the most significant scientific and technological discovery ever made' – that they seemed to think all they needed to do was to convince everyone else. Monsanto's managers apparently assumed that their technologies would be readily adopted by a compliant, global public – they seem to have been taken completely by surprise at the backlash against GM in Europe in the late 1990s.

With Europe's markets effectively shut to them, Monsanto urgently needed to find other consumers, and their attention became even more focused on developing countries. They bought up biotech and seed companies in the Global South, made a public pledge to support poor farmers and to protect the environment, set up a Smallholder Programme, and poured money into research looking at the impacts of GM crops in poorer countries. It's all too easy to stand back and assume that this was just a public-relations exercise, designed to dismantle opposition to the technology, but Monsanto's bosses had been speaking in these pro-poor terms before the European backlash. It seems counter-intuitive, but perhaps the firestorm of controversy and opposition had done something positive for Monsanto itself – pushing its corporate direction closer to the vision of its most evangelical scientists, down a genuinely more humanitarian route. And while it's still easy to be cynical, there's a real possibility that some applications of GM could – as Mike McGrew at the Roslin Institute believed – end up helping some of the poorest communities in the world.

At the same time that Monsanto was pledging to support poor farmers, it was also being generous with its intellectual property. The company freely shared knowledge and technology with public-sector scientists working on the rice genome – including scientists in Europe, working on developing Golden Rice itself.

A golden future for Golden Rice?

The first version of Golden Rice, developed by a team led by Dr Ingo Potrykus of the Swiss Federal Institute of Technology and Dr Peter Beyer of the University of Freiburg in Germany, was unveiled in 1999. Golden Rice graced the cover of *Time* magazine in 2000, but ten years later it was still not available to farmers. Instead, the most common GM crop at the time was herbicide-resistant soy, followed by herbicide-resistant and insect-resistant varieties of maize – all industrial-scale commodity crops. Work on the explicitly pro-poor GM rice seemed to grind on at a much slower pace.

The geneticists working on the original Golden Rice had successfully transferred just two genes – a daffodil gene and a bacterial gene – into a variety of rice, making the plant synthesise its own beta-carotene. In 2005, further genetic tinkering (by Monsanto's main competitor, the Swiss agrochemical and biotech giant, Syngenta) led to the daffodil gene being swapped for a maize gene. The resultant second-generation Golden Rice produced even more beta-carotene than its predecessor.

The originators of Golden Rice chose to stick the new genes into a variety of rice known as *Oryza sativa japonica*, whereas *Oryza sativa indica* is the most widely grown variety in Asia. In order to transfer the 'golden' trait from genetically modified *japonica* into *indica* strains, rice breeders have resorted to conventional breeding techniques. After US field trials in 2004 and 2005, a small-scale trial was carried out in Asia in 2008, followed by more widespread trials in 2013. Agricultural researchers in India are still working on breeding the trait into popular Indian rice varieties. But as of 2016 – there was *still* no Golden Rice seed of any kind available for farmers to grow. The translation of what looked like such a promising advance in the lab, into a real-world crop, has proven to be much more difficult than expected. One sticking point has been that breeding the golden trait into other rice varieties led to reduced yield. But proponents of Golden Rice are also keen to blame the anti-GM movement for slow progress – and there's no doubt that the development of the crop has been held back by both indirect and direct action. Trial crops in the Philippines have been vandalised – not by farmers, but by activists.

As we've seen, some of the antipathy to GM crops – Golden Rice included – emerges from concerns about big science, big business, industrial agriculture and the inability of governments to recognise risks and protect us and the environment. The perceived risks range from concerns over food safety to environmental ramifications, and loss of sovereignty for farmers. The first of these seems easy to dispel: there is no evidence that GM food presents any sort of threat to human health.

The second risk, however, is very real. Wild species are highly likely to become 'contaminated' with genes from GM crops, and it's difficult to predict what the ecological impact will be. In Mexico, the flow of trans-genes from GM maize into old, local varieties has prompted huge concern. In China, where the first GM crops were planted, insect-resistant cotton has been, largely, a success story. But in fact this seems to have happened precisely because regulations were flouted, with the GM trait being bred into local varieties, 'under the radar'. Once this new technology is released, there's no way of putting it back in the bottle.

How we decide to tackle this issue of modified genes escaping into the environment depends a lot on whether GM is seen as merely an extension of traditional breeding techniques – accompanied by interbreeding that has always happened between domesticated species and their wild counterparts – or as an entirely new phenomenon. Proponents of GM tend to espouse the former view, downplaying concerns about switching genes between different species, and encouraging a view of GM as a natural progression within the world of plant-breeding. It's been pointed out that this is rather like saying that the textile mills of the Industrial Revolution were a natural extension of simple spinning and weaving. Nevertheless, other high-tech approaches to producing new crops – such as radiation breeding – have not excited the same level of concern.

The anti-GM lobby is clear that this technology is a game-changer, and radically alters the relationship between humans, their domesticates, and the rest of the natural world. Without sitting on the fence, both sides may have a point. GM fundamentally changes the game – or, at least, significantly bends the rules when it comes to plant-breeding. But then, agriculture – and even hunting and gathering before it – has always impacted on the natural world. It's practically impossible to predict what the long-term effects of this new development will be. This is always a problem with emerging new technologies, and perhaps one of the foremost reasons why governments have been very cautious about allowing GM crops to be planted, applying the precautionary principle.

The third major area of concern – that of food sovereignty in impoverished communities – is also a serious issue. While scientists, politicians and journalists often champion GM technology as 'pro-poor', the evidence for any real benefit to communities in the developing world thus far has been thin on the ground. Most of the currently available transgenic crops are designed and destined for industrialised farms in rich countries. Where studies have been carried out, they generally show GM crops bringing positive economic benefits to poorer countries – but the devil is in the detail. Just because such a crop is grown in a developing country doesn't mean it's grown by poor farmers on small farms. Most of the GM crops in Argentina, for example, are purely cash crops, grown on massive, industrialised farms – they're more about generating profit than feeding a local community.

Nevertheless, GM crops are gaining a foothold in some places. Despite the risks, both real and perceived, it's remarkable how quickly GM crops can be adopted once bans are lifted. In 2001, South Africa legalised the planting of GM white maize; less than ten years later, more than 70 per cent of all white maize grown there was GM. In 2002, Indian farmers were legally permitted to plant GM, insect-resistant cotton; twelve years later, 90 per cent of cotton grown in the country was GM. In 2003, the government of Brazil legalised GM soy; eight years later, over 80 per cent of the country's soy production was GM. Similar, rapid expansions of GM crops happened after the legalisation of yellow maize in the Philippines, GM papaya in China and GM cotton in Burkina Faso. If the problems with yield in new varieties can be overcome, if it works economically, the future for Golden Rice should be rosy. And yet there's one thing that sets it apart from most of these other GM success stories, which could blight its potential – Golden Rice is a food crop.

Perceptions of risks and benefits are very different for an industrial crop – such as maize for animal feed or cotton for the textile industry – compared with a food crop. It's fascinating that, while Europe has been operating a *de facto* moratorium on GM foods for humans, with barriers at government, distributor and consumer levels, there's plenty of GM maize and soybean being fed to animals. Nearly 90 per cent of animal feed in Europe is GM – imported from the Americas. GM *food* has to be labelled as such, but there's never been any requirement to label food products coming from animals that have consumed GM *feed*.

When it comes to food crops, it seems that unfounded anxieties about human health can outweigh more robust evidence of potential benefits to farmers and the economy. In 2002, the Indian government approved the planting of GM, insect-resistant cotton, but in 2009 it banned the use of a GM, insect-resistant aubergine, known as Bt Brinjal. The genetic trait in the aubergine was precisely the same as that in the cotton, based on the insertion of a single bacterial gene. The product of this gene was toxic to insect larvae, and opposition to Bt Brinjal centred around concerns – with no scientific basis – that the insecticidal protein would also be poisonous to humans. Despite protestations from scientists in India and around the world, the Indian minister of environment stuck to his guns and shot down the GM aubergine. It does sound messy. But it's not always the same story. From country to country, from crop to crop, the political, social and economic milieu shifts. In 2013, Bangladesh legalised the planting of Bt Brinjal. So far, the results are looking promising, with reduction in pesticide use and improved yields. But the controversy continues.

Studies suggest that consumers may change their minds if the benefits of a GM food are more obvious to them. An experiment where conventional, organic and spray-free GM fruit was offered at roadside fruit stalls, in New Zealand, Sweden, Belgium, Germany, France and the UK, found that consumers were willing to buy the GM fruit – if the price was right. If the GM fruit presented a pesticide-free alternative, and came in cheaper than organic, it became a palatable option.

If it turns out that there are clear benefits to productivity, economies and human health from incorporating GM crops, such as Bt Brinjal or Golden Rice, into our agriculture – then these need to be weighed up carefully against the risks: trans-genes will inevitably escape into the environment, and there will be social implications too. But vociferous opponents of GM, largely based in the richer countries of the world, need to prudently consider how their opposition affects the ability of farmers in developing countries to make up their own minds about growing GM crops. As political scientists Ronald Herring and Robert Paarlberg put it, 'Farmers in most developing countries will remain unable to use [these] new varieties of food crops ... until consumers in rich countries change their minds about GMOs. Not for the first time in history, the tastes of the rich will drive welfare outcomes for the poor.'

The ill-fated attempt by Monsanto to introduce GM soy into Europe, in the wake of the BSE scandal, acted as a lightning rod for opposition to GM – just as Peter Melchett of Greenpeace predicted. Nearly two decades on, we're starting to understand what the real impacts of GM crops might be. Only time will tell whether Golden Rice will be accepted and take root. It's likely to be available to farmers soon, and its promise – to be a cheap and effective way of combating vitamin A deficiency, as its developers always hoped – will be tested.

And then we'll finally know if it was worth the wait.

Humble origins of a global super-crop

Today, rice is *everywhere*. It was the obvious crop to stick a new gene into if you wanted to combat a globally pervasive vitamin deficiency – placing rice right at the heart of the GM debate. But the origin of domesticated rice itself has also been beset with controversy.

There are two species of domestic rice. *Oryza glaberrima*, or African rice, is grown in a small region of West Africa, and is also a rare crop in South America. Asian rice, *Oryza sativa*, is much more widespread. It includes two main subspecies, *Oryza sativa japonica* and *Oryza sativa indica*. The *japonica* variety, with its sticky, short grains, is essentially an upland plant, grown in dry fields. *Indica* is different – it has non-sticky, long grains, and thrives in lowland, submerged fields – such as Liao Jongpu's flooded, serpentine terraces. Whereas *indica* is almost exclusively tropical, *japonica* exists in both tropical and temperate forms. Both varieties are closely related to the wild rice species, *Oryza rufipogon*. Was one the ancestor of the other, or had they emerged from separate origins?

The wild ancestor of rice, *Oryza rufipogon*, is a wetland plant, growing across a great swathe of Asia, from eastern India, through south-east Asia – including Vietnam, Thailand, Malaysia and Indonesia – to southern and east China. But archaeological and botanical clues pointed to a specific area in this range as the original homeland of domesticated rice – in China itself. This centre of domestication had also given the world domesticated soybeans, adzuki beans, foxtail millet, citrus fruits, melons, cucumbers, almonds, mangoes and tea. The earliest archaeological evidence for crop domestication – with rice amongst those very early domesticates – goes back to around 10,000 years ago.

In 2000, geneticists were bringing their evidence to bear on the question of rice origins, and the archaeological evidence and genetic markers appeared to be telling the same story, of a single origin of *Oryza sativa indica*, in southern China, with *japonica* developing as a later, upland adaptation. But not everyone agreed. Some geneticists argued that the differences between *indica* and *japonica* were too great to have evolved in such a short time span – suggesting that the two varieties had been domesticated independently. Later studies supported that dual-origin model – but there was a hitch: some regions of the genomes of the two subspecies seemed to be more similar than they should have been. And these regions were associated with key domestication traits including reduced shattering, a tendency to grow more upright, with fewer straggling side branches, and a change from a black to a white seed hull. If *japonica* and *indica* came from separate origins, from two different subspecies of wild rice – these genes should not have been the same.

The unrolling of the story seemed to follow a familiar trajectory: early genetic studies, looking at just a few markers, led to the proposal of a simple, single origin; then wider genetic studies came along to suggest multiple, multiregional origins; and then, various parts of the genome seemed to provide conflicting evidence of past events.

In 2012, Chinese geneticists tackled the problem again, publishing their finding in the journal *Nature*. They had carried out genome-wide studies in a range of wild and cultivated rice varieties. They found again that some regions of the genome, particularly those associated with domestication traits, suggested a recent divergence – and therefore a single origin for cultivated rice. However, other regions revealed a much deeper, ancestral story, pointing to multiple origins. It was the closeness of the cultivars to different varieties of wild rice in separate geographic areas that allowed the geneticists to resolve the puzzle. At fifty-five positions in the genome, closely associated with domestication traits, both *indica* and *japonica* were most similar to one particular group of wild rice, from southern China: the ancestors of that wild rice were also the ancestors of domesticated rice. But across the genome as a whole, while *japonica* still looked closest to the southern Chinese wild rice, *indica* was closer to south-east and south Asian wild rices. This makes sense if rice was first domesticated in southern China, as *japonica* then spread westwards, interbreeding extensively with local varieties of wild rice as it went. Of course, rice wasn't migrating on its own – just as in the Near East, the Neolithic in China sparked a population expansion and farmers were on the move. The Y chromosomes of modern Tibetans contain evidence of a wave of migration coming in, between 7,000 and 10,000 years ago. Eventually, domesticated *japonica* rice from the east would come into contact with almost-domesticated *indica*. Once again, then – just as with maize – the story appears to be one of a single origin, and then a spread, involving hybridisation with other wild varieties, or other 'proto-domesticates', along the way.

Musing on the origins of domesticated rice, I can't help thinking back to those handfuls of deeply unpromising-looking seedlings – each just a few blades of grass with roots attached – that Liao

Jongpu gave me to plant in his narrow, twisting, flooded paddy fields. How did this grass species become such an important ally? Just as with wheat, and indeed maize, there seems to be a bit of a mystery around the initial use of wild rice as a food. Before domestication gets to work, before that important non-shattering rachis, before the growth in grain size and yield, it really is difficult to imagine why anyone would turn to this humble grass, with its spray of hard little grains, for sustenance.

Part of the answer lies in the complexity of diets and the drawn-out process of domestication. Although rice seems so important, from a modern perspective, it was actually only a minor crop to begin with. Foxtail millet was more important as an early cereal – domesticated as far back as 10,000 years ago – and its spread seems to have pre-empted the spread of rice. In some ways, though, millet makes rice domestication even more surprising. The wild version of foxtail millet, let alone its domesticated counterpart, boasts an impressively loaded seed-head – something I can imagine a hunter-gatherer being drawn to. It's much less easy to understand why anyone gave rice a chance. But rice didn't suddenly leap from being an unpromising wild grass to being a significant staple. To begin with, it made up just a tiny part of the range of foods that people in southern China were gathering and eating. Early farmers in East Asia cultivated a wide variety of crops, including starchy roots and tubers like yams and taro, as well as non-food plants like gourds or jute. And, as shown at the 8,000-year-old site of Jiahu, near the Yellow River in modern-day Henan Province, they were eating plenty of wild foods too, such as lotus, water chestnut and fish. But – rice was in the mix, and it would gradually grow in importance.

When it comes to the question of exactly where rice domestication first germinated, it seems that archaeologists and geneticists broadly agree – but only broadly. Both genetic and archaeological paths lead back to southern China. But this is a pretty large place. The Chinese geneticists who first published that single-origin-with-later-interbreeding model identified the middle of the Pearl River Valley, in modern-day Guangxi Province, as the homeland of domesticated rice. The impression of timelessness, or at least, deep time, in those famous rice fields of Longsheng could be more than just a romantic notion. (Was Liao Jongpu, I wondered, a direct descendant of the earliest rice farmers? In fact, he's very likely to be – such is the nature of our infolding family trees when you go back that many generations – along with everyone else in China.)

The problem with this genetic identification of the Pearl River Valley as the homeland of domesticated rice is that it clashes with the archaeological evidence – the oldest traces of domesticated rice have been found around the Yangtze River, further north. Around the lower reaches of this river, going back between 10,000 and 12,000 years ago, there's evidence of a growing focus on gathering wild rice, following on from a more sporadic use even earlier. Grinding slabs and wild rice husks have been found in caves and rock shelters in the Yangtze River Valley, dating to more than 10,000 years ago. And then the husks of what appear to be domesticated rice grains have been found in Neolithic pottery at a site in Shangshan, in Zhejiang Province – apparently mixed into the clay to temper it. The pottery dates to around 10,000 years ago. Rice spikelets from the nearby Huxi site, dating to 9,000 years ago, show clear signs of a non-shattering trait – a hallmark of domestication. Rice phytoliths, which vary between wild and domestic rice, have also been used to show a gradual transformation of rice into a domesticate, starting around 10,000 years ago. By 8,000 years ago, several archaeological sites in the Yangtze Valley contain evidence of domesticated rice – based on features of the grains themselves. Then, around 7,000 years ago, the balance starts to shift, with domesticated types starting to outnumber the wild types.

There's always the possibility, of course, that the earlier signal from the Yangtze is just an artefact – that researchers have simply been looking there harder, for longer, or have even been luckier in that area – and that earlier sites from the Pearl River Valley lie, as yet, undiscovered. So rather than relying on a handful of sites, one group of archaeologists took a more sophisticated approach and produced computer models of the spread of rice, using as much of the available archaeological evidence from across Asia as possible. These models also predicted an origin – or, even more likely, two, closely connected origins – of rice domestication in the region of the Middle to Lower Yangtze. If I had to make a bet right now, despite my romantic attachment to that wonderful landscape of Longsheng, I'd go for the Yangtze too.

Winter is coming

The timing of the beginning of rice domestication is important. At the same moment, right at the other end of Asia, people were also starting to cultivate the wild cereals that grew there – rye, barley, oats and wheat. Between 11,000 and 8,000 years ago, those cereals of the Fertile Crescent become staples – and were transformed from wild grasses into domesticates – just as millet and rice were in the Far East.

It seems too much of a coincidence: two groups of hunter-gatherers at opposite ends of Asia developing a novel predilection for wild grasses, becoming increasingly dependent on them, and eventually cultivating them as crops. There's surely something which links these identical changes in human behaviour – something which is at work in the Fertile Crescent, and in the Yangtze Valley, more than four thousand miles away. That 'something' is most likely to be climate change.

During the cold, dry peak of the last Ice Age, wild rice would have been restricted to wetter refugia, in tropical areas of East Asia. As the climate warmed, from around 15,000 years ago, wild rice would have spread, getting an extra boost from the increasing level of carbon dioxide in the atmosphere. Dense stands of wild cereals, replete with grains, presented hunter-gatherers right across Asia with a dependable and easily harvestable food. Under these favourable climatic conditions, wild rice – and millet – may have been a more attractive prospect than it seems today. Perhaps, as we suspect with maize, there were some plants whose characteristics were closer to those which would be bred into whole populations during the process of domestication – plants with larger grains and fewer side-branches – that already looked like a good source of food, and which were easier to gather.

But around 12,900 years ago, the Younger Dryas hit – the cold drought that lasted more than a thousand years. Faced with a dwindling supply of wild food, people may have been driven to control those resources, to start cultivating the wild grasses they'd become so dependent on. The population boom that occurred just before the

Younger Dryas would have meant that resources would have been placed under even more pressure during the climatic downturn. For wheat in western Asia and rice in East Asia – and possibly even for maize in Mesoamerica – the Younger Dryas may have been the crucial factor pushing these species together with humans, forging an alliance which would resonate down the centuries and millennia. As a dependable resource, cereals became more important in diets – becoming a staple. Cultivation would be the next step.

It's a very different view of human history than perhaps we're used to. Not a series of triumphant advances, driven by sheer ingenuity and inventiveness, but a story of misfortune and accidents, contingency and serendipity. People falling on hard times and being forced to change their lifestyles, to adapt and accommodate to a changing environment. The circumstances which led to cereals becoming staples, and then being cultivated, at opposite ends of Asia, makes more sense if we see it as less of a choice – more of a necessity – driven by a climatic downturn.

Yet even if the development of cereal cultivation in western Asia and East Asia was somehow motivated by climate change, the Neolithic evolved in very different ways at each end of this vast continent. In the west, the invention of farming preceded the invention of pottery, with a long 'pre-pottery Neolithic' lasting from nearly 12,000 years ago to around 8,000 years ago. In East Asia, pottery came first, appearing in the archaeological record much, much earlier than the earliest evidence of agriculture. Instead of a pre-pottery Neolithic, there's pre-Neolithic pottery. The dates keep getting pushed back and back.

Pottery from the sophisticated hunter-gatherer Jomon culture of Japan, dating back to nearly 13,000 years ago, was long thought to represent the earliest pottery anywhere in the world. But in the last ten years, evidence of even earlier pottery traditions has emerged in Asia. Sites in eastern Russia and in Siberia have been found to contain evidence of pottery going back an astonishing 14,000 to 16,000 years ago. Analysis of potsherds and associated remains from a cave in Daoxian County, southern China, emerged to suggest a startlingly early date – 15,000 to 18,000 years ago. That study was published in

2009. And then it got even earlier. In 2012, a paper in the journal *Science* announced the discovery of pieces of pottery from Xianrendong Cave in Jiangxi Province – dated to around 20,000 years ago, right at the last glacial maximum. In China, then, pottery was being used some 10,000 years before the development of agriculture. What were the pots used for? Bones from deer and wild boar were also found in the cave – as well as rice phytoliths. Even this far back, in the Ice Age, it seems that hunter-gatherers were eating a bit of wild rice alongside other plant foods and meat. No residue analysis from the potsherds has been reported yet, but black scorch-marks on the outer surface of the pot fragments suggest that they would have been used over a fire. Even if we don't know what they were having for their dinner, it certainly looks like the foragers of Jiangxi were cooking *something* in their pots. The archaeologists reporting on these early pieces of pot talk about the energy gains to be had from cooking starchy foods and meat. But I think that sometimes we focus so much on quite abstract ideas like this that we miss a more obvious benefit. In the depths of the Ice Age, hot food must have been something wonderful to look forward to, after a hard, cold day of hunting and gathering.

Other prehistoric pottery has been shown to be used for storage, food preparation (don't forget the cheese-makers), and brewing up alcoholic drinks. This is a technology which, in China, predates the development of agriculture, and perhaps even helps to push society in that direction – towards complexity, stratification and a more settled way of life. The details of the story vary, but perhaps, once again, we're seeing the old idea of agriculture driving the development of complex society being stood on its head. But we have to be careful – the arrival of settled societies and agriculture comes a long way after the first evidence of pottery use in China. Thousands and thousands of years stretch in between.

Still, the old 'Neolithic package' of pottery, sedentism and farming is smashed into tiny pieces with the discovery of the Xianrendong sherds. By the time people are living in villages like that at the Shangshan site, tempering their pottery with rice, they've made that transition to a settled way of life, and they are cultivating as well as gathering. But the pot-makers of Xianrendong, twenty thousand years ago, were nomadic hunter-gatherers.

The march of rice

Shangshan and other early Neolithic sites in China – dating to around 9,000 years ago (7000 BCE) – provide us with glimpses of the new way of life that would transform people, the landscape, and rice itself. These ancient villages consisted of clusters of rectangular houses, some up to 14 metres long. People were still using old-fashioned Stone Age tools – mostly flakes knocked off stone cobbles – but they also had adzes for hoeing the ground, axes for chopping down trees and grinding stones for pulverising seeds. They were using pottery for storing, preparing and cooking food. They were still, in the main, hunters, gatherers and fishers, but rice would become more and more important.

By 6,000 years ago (4000 BCE), rice was being farmed, alongside millet, across a broad sweep of land between the Yangtze River to the south and the Yellow River to the north. Rice cultivation continued to spread southwards, with extensive cultivation evident in the Pearl River Valley between 5,000 and 4,000 years ago. Rice cultivation also spread north in China, and to Korea and Japan. Early rice cultivation took off in Japan from around 4,000 years ago – this is when rice-seed impressions appear in Jomon pottery. Rice was probably quite a minor crop at this time, being grown in relatively small quantities alongside more important crops like millet and beans – but of course we know that it would only increase in prominence. It's now hard to imagine Japanese cuisine without it.

There's early evidence of rice use in northern India – which prompted archaeologists to suggest the existence of a separate centre of domestication there. Charred rice grains from the site of Lahuradeva in the Ganges region have been dated to around 8,000 years ago (6000 BCE) – but these now appear to be wild rice. It can be a tricky distinction to make, but wild grains tend to have a smooth-edged, circular abscission scar where they separated from the rachis, while domesticated rice grains tend to have a slightly ragged, kidney-shaped scar. By 4,000 years ago, there's definitive evidence of firmly domesticated rice, in the form of rice spikelets from the Neolithic site of Mahagara in north-eastern India. These spikelets had clearly developed a non-shattering trait. This is also precisely when *japonica* was arriving from the east – bringing its domestication genes with it. Other East Asian crops, such as apricots, peaches and cannabis, and stone harvesting knives, similar to those found at older sites in China, also make their way into northern India at this time. Archaeologists suggest that these novelties arrived in India via networks of exchange – precursors of the Silk Road – connecting the cultures of East and South Asia.

It's been argued that, when *japonica* arrived from the east, it was interbred with early Indian cultivars – which hadn't quite developed the full suite of characteristics expected of a domesticated variety. Then interbreeding between the immigrant domesticate and local proto-domesticates could have produced crops which combined the beneficial traits of domestication with local adaptations to climate – becoming *Oryza sativa indica*. But recent research at archaeological sites in north-west India has challenged this sequence of events. At the 4,500-year-old sites of Madsudpur I and VII, 10 per cent of rice grains appear to be domesticated. This seems to be too early for the domestication traits to have been bred in from *japonica* arriving from the east. Archaeologists have suggested that this raises the possibility that there was indeed an independent, though later, centre of rice domestication in northern India. But this doesn't fit with the genetic data. Crucially, the domestication alleles in today's *indica* rice, which are related to the non-shattering trait as well as to a white-coloured husk, and large grain size, have all come from *japonica*.

There seem to be two possibilities here. Either an early variety of *Oryza sativa indica* had already evolved, with its own domestication alleles, that would later be entirely replaced by *japonica* alleles, producing the signature we see in rice today. Or – and perhaps more likely – *Oryza sativa japonica* arrived somewhat earlier than 4,000 years ago in northern India. The only way of settling this question would be to analyse ancient DNA preserved (if indeed there is any) in the grains of rice from Madsudpur.

This debate also focuses on that important difference between cultivation and domestication. Cultivation is something people do to plants – sowing them, tending them, harvesting them. Domestication describes the genetic and phenotypic changes that occur in species when they are under particular selection pressures which are knowingly or unknowingly produced by humans interacting with that species. Even if the rice of northern India was not a true domesticate until it came into contact with incoming eastern varieties, northern India may yet have been, effectively, an independent centre of agriculture. This is certainly true for other crops: there's good evidence that local plants such as mung bean and some small-seeded grasses were cultivated on the Ganges plain long before any crops were brought in from elsewhere. Whatever happened right at the start of rice cultivation in India, by the first millennium BCE the crop was being grown right across the subcontinent.

There's no such argument about the origin of domesticated rice in West Africa. There, an entirely separate centre of agriculture saw rice being domesticated around 3,000 years ago (1000 BCE) – from a completely different wild progenitor. The Neolithic in West Africa had begun with the introduction of cattle, sheep and goats, and gradually the herders had settled in the landscape, and started cultivating cereals such as rice, sorghum and pearl millet, as well as yams. Early farmers around the Niger River cultivated wild *Oryza barthii*, which evolved into the domesticated species, *Oryza glaberrima*, also known as African rice. Genome-wide analyses of African rice suggest that it derived from a single, fairly discrete origin, rather than many separate centres of domestication. These studies also revealed something fascinating about the process of domestication. Geneticists scrutinised the genome of wild and domesticated African rice to look for regions which had been influenced by artificial selection – presumably linked to phenotypic traits which were being selected for. They were interested to see how those traits and genes would compare with those selected for in Asian rice. They looked at homologous (or equivalent) genes in each species – these are genes which are very similar, inherited from a common ancestor of African and Asian rice long before domestication. And what they found were important changes in several genes relating to domestication traits – including husk colour, shattering and flowering. But the changes themselves varied between these domesticated species. For instance, in a certain gene controlling shattering, domesticated African rice showed up a missing section of DNA compared with its wild ancestor. In the equivalent gene in domesticated Asian rice, there was an extra length of DNA compared with wild Asian rice. The effect of these completely different changes to the genetic code – deleting a piece of the code in African rice and inserting a piece in Asian rice – was the same. Both altered genes were linked to reduced shattering. So the Asian and African farmers had each selected for similar traits in the rice they'd grown, and that selection pressure had led to structurally different, but functionally similar, changes in homologous genes. This was good evidence, not only for similar traits being favoured by early rice farmers in both Africa and Asia, but also for the completely separate domestication of African rice. Unlike *Oryza sativa indica*, with its domestication alleles bred into it from *Oryza sativa japonica*, African rice – *Oryza glaberrima* – had its very own, quite distinct, domestication genes.

Wet feet and dry fields

While many plants abhor waterlogged soil, others thrive when fields are flooded. Rice happens to love it, and this secret was discovered back in the Neolithic. The first evidence of wet paddy fields comes from the lower Yangtze Valley, where ancient irrigation systems have been found, dating to the third millennium BCE. There are botanical clues too: archaeologists sifting through the ancient sediments from the Neolithic site of Baligang, situated on a tributary of the Yangtze, have found seeds from wetland weeds, as well as sponge spicules and diatoms – tiny algae with silica cell walls – which all speak of well-watered fields, some 4000 to 5000 years ago. The practice spread, and many archaeologists believe that the beginning of wet-field rice farming in Korea and Japan, around 2,800 years ago, reflects an incoming migration of early farmers.

Inundated fields would have brought key advantages – suppressing weeds and increasing the productivity of rice. How did people first discover this secret? I imagine the way that most discoveries are made – by accident. Perhaps a particularly wet year led to flooded fields – the farmers must have been distraught ... but then the harvest was extraordinary. Once this secret had been discovered, it quickly spread. And the evidence for rice culture eventually emerges from written history, as well as archaeological finds. *The Book of Poetry*, thought to have been written in the eighth century BCE, refers to rice fields irrigated by water diverted from the Shensi River. In the second century BCE, the Chinese historian Sima Qian wrote that the fields of the Yangtze Valley were 'tilled by fire and hoed by water', presumably referring to the use of fire to clear land for agriculture, and the practice of creating flooded paddy fields to suppress weeds.

Cultivated in wet or dry fields, rice was clearly proving to be a useful cereal, and it was spreading. Again, it's easy to fall into that familiar academic trap of talking about everything in the abstract. People didn't start to grow and eat rice because it was a good source of calories, protein and other nutrients. Surely they started to eat it because it was tasty. I love watching cookery programmes, looking at and learning from culinary cultures from around the world. Let's not underestimate our Neolithic ancestors – they had their own cuisines. They would have enjoyed working out how to combine different ingredients to make something new and even tastier. They would surely have jumped at the chance to incorporate something novel into their diets. And if it turned out to be a good, dependable food supply, so much the better. And there lies the secret of any successful ally – combining attractiveness and usefulness.

By the first millennium BCE, *japonica* rice was being grown in tropical south-east Asia, with *indica* rice arriving there later. During the latter part of that millennium, domesticated rice was also spreading west, via overland routes. The traders and armies of the Persian Empire, and then Alexander the Great's Macedonian Empire, helped to introduce rice to the eastern Mediterranean. Carbonised rice grains have been discovered in the pyramids.

But the introduction of rice into Europe – particularly Spain – remains obscure and controversial. Did it arrive by a spread along the northern Mediterranean coasts? Or was it short-circuited – by an introduction across the sea from North Africa? Some claim that rice was already being grown around Valencia during the first century CE. Others suggest that the Moors (from the North African lands known by the Romans as Mauretania) introduced rice – along with saffron, cinnamon and nutmeg – to Spain, much later, in the seventh century CE. After all, the Spanish word for rice, *arroz*, comes from the Arabic *al arruz*.

However it was introduced to Spain, rice was regarded by other western Europeans as a food for infants. And yet the Spanish embraced it, recognising its culinary potential and laying the foundation for one of the most well-known Spanish dishes – paella. From Spain, rice cultivation spread to Portugal, and to Italy, between the thirteenth and fifteenth centuries CE. Today, Italy and Spain remain the largest rice producers in Europe.

After Columbus's voyage of discovery, domesticated rice became part of the great Atlantic exchange, crossing from the Old World to the New. For Latin Americans living in tropical countries today, rice is the single most important source of calories after sugar. The combination of rice and beans is so iconic, so important in Caribbean cuisine in particular – and yet it's a relatively recent partnership. It dates back just a few hundred years, and has been called 'an early dish of globalisation'. But the underlying concept – the idea of combining grass seeds and pulses – has an ancient heritage, going back before the origins of farming. These foods may complement each other in taste and texture, but they do something even more important than that: they make up for each other's deficiencies. They come together to create a comprehensive package of protein that includes all the amino acids – the building blocks of proteins – that the human body needs but cannot make.

In each centre of domestication – including East Asia, the Fertile Crescent, West Africa, Mesoamerica and the Andes – early farmers domesticated at least one indigenous species of grass and one indigenous species of legume or pulse. Today, the descendants of those founder crops of cereals and legumes feed the majority of the global population. In the Fertile Crescent, early farmers grew lentils, peas, chickpeas and bitter vetch alongside emmer wheat, einkorn and barley. The farmers of the Yangtze Valley were growing soybeans and adzuki beans alongside rice and millet. The separate centre of agriculture in sub-Saharan West Africa saw the cultivation and domestication of hyacinth beans and cowpeas alongside pearl millet, finger millet and sorghum, between 5,000 and 3,000 years ago. In the Americas, common beans (also known as string beans, and – quite wrongly – as French beans) and lima beans were grown alongside maize.

The great Atlantic exchange saw crops swapped between Old and New Worlds, and centuries of slavery left their mark on agriculture too. Spanish settlers arrived in the Americas with rice to plant there – as a subsistence crop. The Native Americans had gathered and eaten indigenous wild rice, but the Asian rice was softer and more palatable. It grew well in wet lowlands, where maize could not be grown. The immigrant rice would develop into a staple crop in Latin America and the Caribbean. By the eighteenth century, rice was being grown on a huge scale in South Carolina, mostly for export.

Enslaved Africans brought sorghum and African rice, *Oryza glaberrima*, with them to the New World – though Asian rice, *Oryza sativa*, was higher yielding than its African cousin, and became the predominant crop. So the famous rice and beans of the Caribbean is a truly cosmopolitan dish, most often combining Asian rice with pigeon peas, *Cajanus cajan* – originally domesticated in India and arriving in the Americas via Africa. In this apparently simple food, then, is an astonishing depth of history, stretching from the first farmers of the Yangtze Valley and India, to European contact with the New World and the transatlantic slave trade. The best and the worst of globalisation and human interactions are enshrined in this dish.

The European colonisation of Africa left its mark on crops there too. Some five hundred years ago, Portuguese colonisers introduced Asian rice, *Oryza sativa*, to West Africa, and it largely replaced African rice, thanks to its higher yields. African rice is now grown only on a small scale, as a subsistence crop – but it continues to hold a special cultural significance for some; the Jola people of Senegal grow it specifically for use in rituals. And Asian rice, while outperforming African rice in some ways, fails badly in others. It's not as good as suppressing weeds as its African counterpart, and it is *extremely* thirsty – not really a crop suited to an African climate. And as the population of Africa has risen, rice production has not kept pace. In the 1960s, sub-Saharan Africa produced more rice than it needed; by 2006, it was producing less than 40 per cent of the rice being consumed.

In the 1990s, plant-breeders setting out to produce new varieties of rice, specifically suited to an African environment, worked on hybridising African and Asian rice. The aim was to combine the high-yielding traits of *Oryza sativa* with the drought resistance of *Oryza glaberrima*. The project was called 'New Rice for Africa', or NERICA. The cross the rice-breeders were aiming for was quite tricky – after all, they were trying to bring together two separate and quite distinct species. African and Asian rice do not breed together naturally. So the scientists used a plant version of IVF. The resultant hybrid embryos needed careful support, and were grown in tissue cultures in the lab. But – it worked: thousands of new, hybrid varieties were created, and are already being grown in Guinea, Nigeria, Mali, Benin, Côte d'Ivoire and Uganda. The results – at least, those reported by the NERICA project – look promising: the hybrids produce a higher yield than their parent varieties, contain more protein, and do seem to be more drought-resistant than Asian varieties. But NERICA is not without its detractors – who see this as yet another example of a top-down solution being imposed on poor farmers, without proper engagement. They raise familiar concerns, worried that it could promote monoculture and devalue local seed systems, without fully delivering on its promise.

NERICA hybrid rice brings us back, full circle, to Golden Rice – making us look again at some of the philosophical objections to genetic modification. Creating hybrids by breeding separate species together has long been considered acceptable in agriculture, while moving individual genes, or suites of genes, across the species boundary provokes anxiety.

NERICA also demonstrates how important it is to preserve diversity – even while some species and strains are so successful, they seem set to supplant all the rest. We saw the danger in such a narrow focus with the Lumper potato, its susceptibility to disease – and the famine that ensued. The diversity of our domesticated species, and their wild counterparts, represents a vast storehouse of variation – of adaptations that have proved useful, in different times and different places, under domestication and out in the wild. There's still room to improve existing crops, and that living archive presents us with opportunities to do just this – whether through age-old breeding or new techniques like gene editing. Not only that, but human needs will change, as will climate and environments. Some of the less promising varieties of today may well come into their own in the future – if, that is, they're still around.

But NERICA also reminds us that, no matter how beneficent the intent, and irrespective of the technology used to drive advances in agriculture, scientists and farmers need to work closely together. The potential of advanced agricultural technology to change lives, to save lives, will only be recognised through real engagement – not just with abstract problems, but with the people working the land. People like Liao Jongpu and his predecessors have been preparing the ground, sowing seedlings, reaping the harvest and sharing that bounty with their communities for centuries, for millennia. They're not just 'end-users' – they drive innovation too. There's not only a moral imperative to involve them in development, farmers will help us all to make better decisions. They've been in the business of domestication and crop improvement for thousands of years.

8

HORSES

Equus caballus

O I *was thine, and thou wert mine, and ours the boundless plain,*
Where the winds of the North, my gallant steed, ruffled thy tawny mane ...

William Henry Drummond, 'Strathcona's Horse'

Un caballo llamado Zorrita

Zorrita was my companion for just three days, and we were very close. We were thrown into each other's company, but pretty much immediately we understood one another. For that short time, we looked after each other and I grew extremely fond of her. She became a firm friend. But when I said goodbye, I knew I was unlikely to ever see her again.

There was something of a language barrier on that first day, but I quickly learnt to communicate with Zorrita, and she understood exactly what I wanted. Together we trekked along valleys, through rivers and up mountains. She carried me all the way, taking my direction, but picking the best path herself through prickly bushes and up steep, rocky mountain ridges.

I first met Zorrita at the stables on the Cerro Guido ranch, in the Las Chinas Valley, near the Torres del Paine mountains in southern Chile. I was introduced to this horse by a gaucho named Luis. He was dressed in loose, black linen trousers, tall leather boots, a red shirt and a brown jerkin. He wore a black cap with a red cord around it, his tousled long black hair escaping at the back. His stubbled face and his hands were brown and weathered. I guessed he was around fifty, but he could have been younger. He'd clearly spent most of his life outside and around horses. He hardly spoke any English – and I hardly speak any Spanish – but somehow he asked me if I'd ridden before, and I said yes, a little. Zorrita, he told me, was a special horse. A champion. I was excited and daunted as I swung up into the saddle.

I'd grown up with the English way of riding, holding the reins in two hands, fitting feet securely into stirrups, and lifting out of the saddle to gallop. Western-style horse-riding is quite different – you hold the reins in one fist, with just your toes in the stirrups, and sit firmly into the deep saddle when galloping. I'd had a chance to experience this way of doing things before, but some years ago, and it still felt a little alien to begin with. But I settled in quite quickly – what was surely far more impressive was that Zorrita seemed to immediately understand her new rider. After a few minutes, she was perfectly attuned to what I wanted: where I wanted her to go, and how fast. We left the stables and made our way up into a long valley, with snow-covered mountains in the distance. After an hour of walking and trotting, Luis rode up alongside me.

'*Bien?*' he enquired. '*Muy bien,*' I replied. 'Gall-op?' he asked, and before I had a chance to answer, he'd spurred his own horse into action, and I had little option but to do the same with Zorrita – she'd been wanting to run since we'd left the stables – and we were soon flying down the valley, hooves thundering on the turf. It was utterly exhilarating.

After three hours of riding, we reached our destination and set up camp, down by the river. I was hunting for dinosaur fossils with a Chilean palaeontologist called Marcelo Leppe. His site was high up in the mountains above us, and the following day we rode up to it. The first part of the ascent was steep, but over grassy, mossy terrain. As we climbed higher, the vegetation ended and we were riding up an even steeper, dusty and rocky mountainside. It rose up, practically at a 45-degree angle. I looked at Luis, above and in front of me. His horse was perched, seemingly precariously, on the sheer, stony slope. I followed him up on Zorrita. She seemed a little wary at first, testing her footing (her hoofing?) on the rocks. She picked out a narrow trail of her own devising. There were no real paths up here. Her hooves dislodged a couple of rocks and they went tumbling and skittering off down the hillside. I tried not to look down after them. We turned a corner and found ourselves on a more gentle incline – covered in vegetation again. It had been a false summit – we still had some way to go, to reach the fossil site near the top, but the most precipitous, dangerous-looking part was over. I breathed a sigh of relief. In fact, I think I'd been half-holding my breath for most of that tricky ascent.

We reached the fossil site and spent a fruitful few hours collecting surface finds – laid bare by the winter's snow, now melted, and the wind, which still whipped sand into our faces as we searched for ancient relics. I found a piece of a vertebra of a hadrosaur, a 68-million-year-old duck-billed dinosaur, and several pieces of fossilised monkey puzzle tree – so well preserved that the grain of the wood and even the tree rings were still clear.

Then it was time to go back down to the camp, before darkness enveloped us. The descent was even more terrifying than the journey up the mountainside had been. Now it was impossible not to look down. I stood in the stirrups and leant right back in the saddle. If Zorrita had slipped, both of us would have ended up at the bottom of the slope. I could have dismounted and descended on foot, but I trusted her – and she got me down safely.

What an extraordinary partnership with another being. And it's one that depends on centuries of humans and horses getting to know each other, working out how to communicate, and establishing trust. It also seems to depend on an innate predisposition of horses – something deep within them – that means, like dogs, they can enter into this inter-species partnership. They're naturally gregarious creatures. Wherever we stopped en route or at the camp, Zorrita clearly wanted to be close to the other horses. When we were ready to set off, she would nudge the others, pushing her head against their flanks and shoulders, nuzzling their noses. They'd do the same to her. We'd left a couple of the horses tied up at the camp. As soon as Zorrita spotted them on our return down the mountain, she neighed excitedly. They neighed in reply. They were obviously very pleased to see each other.

The gauchos took the horses back to the stables each evening, and rode back up the valley to our camp each morning. On one of the evenings, we heard that they'd managed to catch one of the wild horses, *los baguales*, that roamed around the Las Chinas Valley. On our last day, we struck camp and rode the horses down the pass. I dismounted in the paddock and tied Zorrita to a fence post, whispering an affectionate farewell to her and patting her shoulder. She stood there calmly as the rest of the expedition arrived, and all the horses were tied up in a line along the fence.

The *bagual* was standing in a corner, tethered away from the others, with a basic rope bridle. His black mane and tail were splendidly long. He looked more curious than frightened, but his life as a creature of the wilderness was at an end. His feral nature would be tamed. He'd be a fine addition to the stables, I could see that. And there, he'd be protected from pumas, and fed plenty of hay. Still, I couldn't help feeling a little sorry for him.

As I walked off and closed the gate behind me, Zorrita had a magnificent tantrum. I like to think it was because I was leaving. She reared up with such force that she pulled the stout fence post clear out of the ground. A raucous rumpus broke out, neighing and flying hooves, but the gauchos ran in and quickly held ropes and soothed the tired horses. Zorrita had been lucky not to injure herself, and she soon calmed down. She was tame enough, but still wild in her heart.

Horses in the New World

The wild horses of Chile seem to belong in that untamed landscape –
as much a part of the wild, natural country as the guanacos, pumas,
armadillos and condors. And yet the ancestors of the *bagual* that the
gauchos caught in the Chinas Valley could only have been there for a
few hundred years. For thousands of years before the Spanish and
Portuguese arrived there had been no horses in the Americas. The
ancestors of the *baguales* were domesticated – they are not truly *wild*
horses, but feral.

And yet, going much further back in time, there were plenty of
horses and earlier, horse-like creatures, roaming the Americas. In fact,
the origin of this group, and many of its numerous branches, was in
North America. The evolutionary history of horses and their ilk
includes great proliferation of an ancient family tree, and diversifica-
tion – as well as a harsh cutting back of many branches, until just a
fraction of the splendid, ancient diversity exists today.

Horses are classified as odd-toed ungulates (hoofed animals). Now,
it's not that their toes are strange, just that they only have one of them,
an odd number. Rhinos and tapirs are also odd-toed ungulates, or
Perissodactyla – but with three toes. The fossil record of the Equidae,
the family which includes modern horses, goes back some 55 million
years – starting with the dog-sized *Eohippus* of North America. These
early equids were still in possession of several toes on each foot –
three-toed on their front feet; four-toed on their hind legs. Over time,
they'd lose them all except one. With plenty of fossils to show the
gradual loss of toes, this classic example of evolutionary change in
anatomy is enshrined in biology textbooks.

At times of low sea level, early horse-like creatures could spill out
of North America, across the Bering land bridge into Eurasia. There
was an early expansion of small, leaf-eating equids out of America
into Asia around 52 million years ago – but the descendants of those
pioneers later died out. The family tree of horses goes crazy in the
Miocene – the geological epoch that lasted from 23 to 5 million years
ago. North America filled up with a huge range of horse-like animals

of different shapes and sizes: some leaf-eating browsers, some grazers – all fleet of foot. By five million years ago, the fossil record of equids incorporates more than a dozen distinct genera – groups of species – of horse-like creatures, including the three-toed *Merychippus, Pliohippus*, the earliest one-toed horse, *Astrohippus* and *Dinohippus* (the ancestor of modern horses), to name just a few. Again, some – like *Sinohippus* and *Hipparion* – spilled out, across Beringia, into Asia.

At the beginning of the Miocene, North and South America were separated by a large body of water, called the Great American Seaway. In the middle of the Miocene, volcanoes at the bottom of the Seaway created a scatter of islands between the Americas. Sediment gradually accumulated around the islands, until eventually the Isthmus of Panama was created. The emergence of this land bridge allowed plants and animals to spread from North to South America, and vice versa. The migrations peaked around 3 million years ago in what has become known as the 'great American interchange'. And as part of it, horses expanded down into South America. The first to arrive belonged to the genus *Hippidion* – a separate, now extinct lineage. They were funny-looking little horses with short legs. By a million years ago, *Hippidion* would be joined – in South America – by true horses, *Equus caballus* – essentially the same species as our domesticated horses today.

The tale of the equid family is one of severe pruning as well as burgeoning proliferation. Of all those diverse genera in the Miocene, only one lineage made it through to the present day: the genus – *Equus* – to which all the living horse-like animals belong, from actual horses (officially: caballines) to asses, donkeys (the domesticated descendants of the African wild ass) and zebras. Geneticists have been able to extract and sequence DNA from a horse bone preserved in permafrost in the Yukon, dating to 700,000 years ago – the oldest genome yet. Based on the differences between that ancient genome and those of modern equids, the geneticists have concluded that the *Equus* lineage originated around 4 to 4.5 million years ago. Then the caballine and the zebra–asses lineages diverge away from each other about 3 million years ago.

Some 2 million years ago, an expansion out of America saw the ancestors of modern asses and zebras arriving in Asia, spreading to Europe and down to Africa. Then, some time after 700,000 years ago, the ancestors of our modern horses also traversed the Bering land bridge from North America into north-east Asia. They quickly expanded across Eurasia. Fossils of two equine species, one an ass, the other an ancient horse, have been found at the early Middle Pleistocene site of Pakefield in Suffolk, dating to at least 450,000 years ago, and at Boxgrove in Sussex, dating to 500,000 years ago.

Having originated in North America, before spreading to South America and to the Old World, *Equus* would ultimately go extinct in its homeland. Some 30,000 years ago, as the ice sheets were descending over North America, the endemic, 'stilt-legged' horses disappeared from the landscape. In South America, *Hippidion* and caballine horses clung on longer, until after the last glacial maximum. If I'd been able to travel back to the Las Chinas Valley, perhaps 15,000 years ago, I may have seen truly wild horses – and perhaps species of both *Equus* and *Hippidion* – there. But they wouldn't be around for much longer. And it wasn't only the climate that was against them.

Around the peak of the last Ice Age, sea level was low and human hunters would have been able to cross Beringia into the northernmost reaches of North America. Butchered horse bones have been found in the Bluefish Caves in the Yukon, dating to around 24,000 years ago. But access to the land further south was blocked by vast ice sheets. By 17,000 years ago, the ice sheets were melting at the edges – enough to allow human colonisers to migrate from Beringia and the north-east tip of North America, down into the rest of the continent. By 14,000 years ago, there's plenty of evidence for human occupation right across North America, and down into South America too. And these humans carried some formidable hunting weapons.

Horse bones turn up occasionally in North American archaeological sites, associated with human occupation or activity. At Wally's Beach, above the St Mary River in south-western Alberta, Canada, wind erosion has helpfully exposed ancient sediments from the very end of the Ice Age. And pressed into the ancient mud are the preserved footprints and trackways of extinct American mammals – this was clearly a well-used game trail. But alongside the tracks of these long-gone animals, there were bones too – of horse, musk oxen, extinct bison and caribou or reindeer. Some of the horse and camel bones had clearly been butchered. The site has also yielded human artefacts, in the form of stone flakes, which were probably the tools used on the carcasses. The evidence at Wally's Beach includes eight separate butchering localities.

Archaeologists have suggested that these localities were almost contemporaneous – it's possible that animals were being butchered at these separate sites during the same year, the same season, possibly even during the same hunting trip. But are they really evidence of hunting, or could those ancient Paleoindians simply have been scavenging carcasses that had been killed by other predators? No hunting weapons were found at the butchering sites themselves, but a few stone points or spearheads were discovered nearby. And when archaeologists tested these points, they found two bearing traces of horse protein.

The stone points – which are beautiful, carefully flaked spearheads – are of the Clovis type. The oldest firm dates for Clovis culture in North America place its emergence around 13,000 years ago. The Wally's Beach stone points were 'out of context' – it's impossible to get a direct date on these spearheads. They were found some distance away from the bones at the butchering site, which were themselves dated to 13,300 years ago. So this leaves two possibilities: either the spearpoints, bearing horse protein, represent slightly later hunting of horses by Clovis people, after 13,000 years ago; or Clovis culture emerged a century or two earlier than previously thought. Whether the finds from Wally's Beach represent at least two events, separated by a few centuries, or a single event, may be a question that remains impossible to resolve. Nevertheless, the points *do* provide unequivocal evidence – the Stone Age equivalent of a smoking gun – of the hunting of horses by ancient people in North America.

The last specimens of *Hippidion* – found in Patagonia – date to 11,000 years ago. Caballine horses in both North and South America may have clung on a little longer – but their days were numbered. The last trace of truly wild horses in North America comes not from bones, but from DNA preserved in sediment in Alaska – dating to 10,500 years ago. The debate rumbles on about whether it was climate or humans that finished off the indigenous American horses. There was an overlap of several thousand years between the arrival of humans in the Americas and the disappearance of the horses. So human hunters certainly weren't rampaging across the land on some sort of violent, overkilling spree. On the other hand, we can be sure that they were hunting these animals, even if only occasionally, and that would have had some impact on the already dwindling population. Although climate and changing environments are probably mostly to blame, humans may have helped to speed the extinction of American horses.

By the nineteenth century, the memory of ancient horses in the Americas had completely faded. As far as everybody was concerned, horses were firmly Old World animals, introduced to the Americas by the Spanish. And then, on 10 October 1833, a ship's naturalist from Britain was exploring the coast near Santa Fe, recording the geology and any fossils he came across. He was investigating a fossil of an extinct, giant armadillo, when he found what appeared to be a horse's tooth in the same layer of reddish sediment. It looked a bit weird compared with modern horses' teeth, but definitely horse-like, nonetheless.

The naturalist – none other than Charles Darwin – pondered in his field notebook as to whether the tooth might have washed down from a much later layer, but he concluded that this was unlikely. The tooth was extremely old. Darwin had found the first evidence of an indigenous, ancient horse in the Americas.

When he returned home, Darwin wrote up his discoveries in the *Journal of researches into the geology and natural history of the various countries visited by H.M.S. Beagle* – the book which would later be rebranded as *The Voyage of the Beagle*. And he returned to that horse's tooth in his *Origin of Species*, writing: 'When I found ... the tooth of a horse embedded with the remains of Mastodon, Megatherium, Toxodon, and other extinct monsters ... I was filled with astonishment.'

The eminent nineteenth-century anatomist Richard Owen (who later became – I think it's fair to say – the closest thing Darwin ever had to an arch-enemy) wrote up the fossil mammal remains collected during the voyage of the *Beagle*. He looked at the tooth from Argentina and had to admit that Darwin was right. He wrote that the tooth '... from the red argillaceous earth of the Pampas at Bajada de Santa Fe ... agreed so closely in colour and condition with the remains of the Mastodon and Toxodon from the same locality, that I have no doubt respecting the contemporaneous existence of the individual horse, of which it once formed part.' Then, he continued, begrudgingly: 'This evidence of the former existence of a genus, which, as regards South America, had become extinct, and has a second time been introduced into that Continent, is not one of the least interesting fruits of Mr. Darwin's palaeontological discoveries.'

It *was* an interesting fruit. No wonder Darwin had been 'filled with astonishment'. It was a real revelation: when the Spanish brought horses with them to the Americas on the cusp of the sixteenth century, they were *reintroducing* a lineage which had existed for thousands of years in the New World – and which had in fact originated there. Darwin went on to use his fossil horse's tooth from Santa Fe to illustrate his ideas about extinction in the *Origin* – proving that ancient horses had once galloped across South America, and then disappeared, long before Columbus made his voyage of discovery.

Horses in the Old World

While horse populations were dwindling, eventually to disappear completely, in the Americas, their relatives – horses, asses and zebras – survived in the Old World. Large herds of wild horses continued to roam across northern Siberia and Europe while their American cousins were facing extinction.

It seems odd that horses went extinct in the Americas at the end of the Pleistocene, whilst surviving in Eurasia. They were facing similar pressures in both places – climate change and human predation. And horses had been feeling the sharp end of human hunting weapons for much longer in Eurasia than in the Americas.

Our own species, *Homo sapiens* – originating in Africa some 300,000 years ago – had expanded into both Europe and Siberia by at least 40,000 years ago. But way before that, horses had been predated by earlier populations of humans, going back hundreds of thousands of years. At Boxgrove in Sussex, a 500,000-year-old horse scapula with spear damage shows that early humans – probably *Homo heidelbergensis* – were hunting horses. At the last glacial maximum, the horse population of north-west Europe would have plummeted, under attack from both the icy conditions and the lethal spears of the Palaeolithic hunters.

The Ice Age inhabitants of western Europe were very familiar with horses, and these animals formed the subjects for some cave paintings – images that would be discovered and wondered at millennia later. At the famous cave of Lascaux, near the town of Montignac in the Vézère Valley, in south-west France, small, pot-bellied horses run along the walls, alongside bulls and reindeer. They're thought to have been painted around 17,000 years ago. My favourite Ice Age painting of horses comes from another cave, however, about 100 kilometres to the south of Lascaux – Pech Merle. The paintings inside this cave are believed to be even more ancient, perhaps around 25,000 years old. I was lucky enough to visit the cave in 2008, with just a few other people, and I wrote about what I saw there:

> *A flight of stone stairs led down ... I passed [through a door] to emerge into a limestone cave deep within the hillside. I walked through magnificent chambers with huge flowstone creations, enormous stalagmites and stalactites, some of which had met between ceiling and floor to form massive pillars. The cave opened into a great chamber ... there, on one rare, smooth part of the cave wall to my left, were two beautiful horses outlined in black, facing away from each other, their hindquarters partly superimposed. They were covered in black spots which also flowed on to the background around them, as though they were somehow camouflaged. There were red ochre spots, too, on the belly of the horse on the left, and on the flanks of the other. I noticed that the flat wall of rock had a strange contour where it ended on the left – almost like a horse's head. It was as though*

the artist had taken this suggestion from the natural shape of the
rocky canvas ... The horses were stylised rather than naturalistic
representations. They had great curving necks and small heads,
rounded bodies and slender legs. Were they artistic representations of
real horses or mythical beasts?

Whatever those images represented – imaginative riffs on real horses, or horse-spirits, or even horse-gods – we can be sure that the Palaeolithic hunter-gatherers of Europe not only knew what horses looked like, they knew what they tasted like, too. There are plenty of Ice Age archaeological sites with butchered horse bones. In fact, horse – together with bison – is the most common large mammal in archaeological assemblages. Around 60 per cent of late Ice Age archaeological sites in Europe and Siberia contain horse bones.

After the peak of the Ice Age, the climate began to improve and the potential range for horses increased, with plenty of pasture rolling out – but their numbers kept falling. The continued pressure on the Eurasian population of horses must surely have been exerted by human hunting. And by this time, of course, the hunters of Siberia and Europe were accompanied by dogs.

The world kept warming up, and the environment kept changing: grasslands were shrinking, as Europe became increasingly forested. The cold snap of the Younger Dryas interrupted the trend, with the forests of western Europe reverting briefly to glacial tundra – but then warmth returned. By 12,000 years ago, the open, Ice Age grasslands known as the 'Mammoth steppe' had all but disappeared from Europe – together with the mammoths themselves. Instead, there was now extensive woodland – mainly birch in the north, pine in the south. From around 10,000 years ago, the central European lowlands were colonised by much denser, mixed deciduous forest, with oak as the predominant species. Warm-loving, forest-dwelling animals such as deer and brown bears were suddenly in their element; they spread north from refugia in southern Europe. Horses, however, were facing habitat loss, and by 8,000 years ago they'd disappeared from central Europe. But there were other areas where much more widespread, suitable habitat existed, on into the middle of the Holocene. These were the grasslands of Iberia, and the Great Steppe or Eurasian Steppe, stretching from north of the Black Sea, as the Pontic-Caspian Steppe, through Russia and Kazakhstan to Mongolia and Manchuria. Plenty of grazing, then, in those grasslands – but also plenty of hunters.

Even in Europe, there appear to have been some refugia – pockets of suitable habitat – where small numbers of horses could cling on. There are over 200 archaeological sites, dating to between 12,000 and 6,000 years ago, ranging from Britain and Scandinavia to Poland, which preserve evidence of wild horses. This suggests that – although the new forests were too dense for animals like mammoths and giant deer, which then faced extinction – there were enough woodland groves for horses to graze in, even if their populations were now small and fragmented. Forest fires – common in pine forests – could have helped to create clearings. Along the course of large rivers, regular flooding could also have kept woodland at bay, creating river meadows suitable for large, grazing mammals.

And there was something else that helped to create habitat for wild horses. Around 7,500 years ago (5500 BCE), the frequency of horse remains found in archaeological sites across Europe increases. This upsurge in horses seems to coincide with the arrival of a new way of life in Europe: farming – and the beginning of the Neolithic. When early farmers began to fell trees to clear space for agriculture, cattle and sheep, they were inadvertently making room for wild horses too.

For species that became our allies, that teamed up with us, benefits could be even more direct and palpable. The end of the Ice Age was a time of great ecological upheaval. Many large-bodied mammals died out between 15,000 and 10,000 years ago – including enormous, iconic herbivores such as mammoths and mastodons. Predators were also hard-hit, as their prey dwindled away. The cave lion disappeared from Eurasia around 14,000 years ago; the American lion went extinct around 13,000 years ago. Sabre-toothed cats clung on in the New World until around 11,000 years ago. The wolf population survived but was still badly hit – though of course one lineage went on to become immensely successful: the wolves that took up hunting with humans, and became dogs. It's estimated that there are well over 500 million dogs in the world and around 300,000 wolves. So today, dogs outnumber their surviving wild relatives by more than 1,500 to 1. No one seems to be willing to hazard a guess at how many red junglefowl there are in the world, but it's bound to pale in comparison to the global chicken population of at least 20 billion – around three chickens to every person. And cattle – well, there are no surviving aurochsen and an estimated one and a half billion cattle on the planet.

Wild horses have led a similarly precarious existence. Loss of habitat and hunting by humans laid low their population. The forest-clearing of the Neolithic might have provided some pockets of habitat, and a temporary boost, but numbers would continue to fall. The population of *Equus ferus* – the close wild relatives of domestic horses – dwindled to nothing in the twentieth century. The last wild horse in Mongolia, belonging to another species, *Equus przewalskii*, was spotted in the 1960s. But then they were reintroduced, and now there are an estimated 300 living wild, with around 1,800 in captivity. In a curious twist of fate – and another unexpected and challenging example of the impact of human activity – it turns out that wild horses, along with moose, deer, boar and wolves, storks, swans and eagles, are doing very well in the exclusion zone around Chernobyl. The positive impact of removing humans from this area seems to be overshadowing the negative effects of the radiation itself.

But – not all horses stayed wild, of course. You've probably never seen a wild horse, but I imagine you've seen plenty of domesticated ones. You may even have ridden one. Did it stand there, meekly, as you swung your leg up and slid into the saddle? Probably.

I wouldn't describe Zorrita as meek, but she was perfectly happy for me to jump up on to her back. She didn't try to throw me off, at any rate. Can you imagine what would have happened had I tried to mount that feral horse, that *bagual*? He would have had none of it – and neither would his wild predecessors. However much we marvel at humans having become close to wolves, which meant trusting that they wouldn't use their strength and their terrifyingly sharp, slicing teeth on us, it's surely just as astonishing to trust yourself to a large, fleet-footed mammal – one which could easily rear up, or career away, and throw you off, doing some very serious damage.

Breaking in

Imagine catching a wild horse – for the very first time. No one you know has ever done this before. And you bring it home, biting and kicking. You tie it up. You feed it. Your family thinks you're mad. They want you to kill it – after all, it would feed everyone for *weeks*. But you want to keep this young, wild thing alive. You like it. And you have an idea. Everyone thinks you're crazy.

You wait until the wild horse gets used to you approaching. You get closer and closer. She lets you stroke her mane, her neck. And then – you grab the mane, and fling yourself up on to her back. She's not happy. She pulls at the rope you've tied to a post. She bucks to get you off her back. You lie down and hold her neck. You cling on. As she calms down, you sit back, releasing your grip around her neck. You hold tight to her mane instead.

After a few moments, while she's snorting and stamping, but not trying to buck you off, you move one hand down the rope around her neck. And you gently loosen the knot. She knows. As the rope drops to the ground, she knows that she's free. You – are just a distraction. She swings around, plunging her hooves into the wet ground, and RUNS. Hooves flying, you hear her breath in rhythm with the footfalls. You are clinging on for dear life. This is like flying and dying and being born and feeling one with the wind and the wildness and the landscape and the heavens. You hold on. The cadence of the gallop knocks the breath out of you as you bounce up and down on her back. She pulls steep turns, trying to knock you off. But you stay on. She runs and runs and runs. You're very far from home.

At long last, she tires. Snorting and throwing back her head, spraying you with horse snot. She's cantering now, her flanks and neck damp with sweat. Your hands are knotted into her mane. She trots, walks, stands still. You both stay still and breathe for a while. The gallop was exhausting, terrifying ... exhilarating.

Then you sit up a little. Pull gently on her mane. You'd like her to turn around. She does. Now you're facing in the direction you'd like to go. The camp is somewhere that way – along this valley, to the left of that hill. Can you ask her to take you back?

You shift your weight forwards a little. She responds by stepping forward. You stroke her mane. You shift forwards again, pressing into her flanks with your feet. She breaks into a trot. You're trying not to grip her neck too tightly. If you can sit back a little, you can pull her mane to one side or the other – you can guide her. You've made an astonishing connection with a wild animal. You splash down into the river and up the opposite bank, round the flank of hill – and you can see the camp, the tents, and the smoke from the fires snaking up into the sky. What are they going to say when they see you riding in on this magnificent creature? You've captured her spirit – felt her power – in a way you'd never do by hunting and killing and eating her. The potential of horses has been unlocked. You feel like a god among people. And they're running out to greet you – your sisters and brothers, your parents, your uncles, aunts, cousins and friends.

You're almost at the camp. Your horse is slowing – she would usually stay away from humans. You urge her on. She's yours now.

One of the camp dogs runs out and sniffs at her legs. She rears up. You try to cling on to her mane. She's rearing then kicking back and flinging herself from side to side. You are thrown off, flying up and then landing on your back. It knocks the air out of you. You lie on the ground, gasping and gasping. You're all right. A pain in your ribs will last a while but will heal and disappear. As you get your breath back, you bring up your left hand, against your chest. Fingers uncurl to reveal a clump of black, wiry horsehairs in your hand. You travelled with her. Now she's gone, but you'll always remember that wild ride.

After that, all your friends want to try. It becomes like a game. Who will dare to catch and ride a horse? It's exhilarating foolishness. The stuff of youth. But before long, there's a small group of you who not only ride horses, you keep them. Together, you become a force to be reckoned with. Headstrong youngsters – but a rising elite.

Years later, when you're an elder of the tribe, and horses are all around, you'll tell the story. *Once they were all wild, these beasts that now seem like our allies*. And you were the *first* person to try the unthinkable, to try riding a wild horse. You broke the spell. Even though you lost that first horse, even though you cracked your ribs as you hit the ground, people saw what was possible. Things have changed so much during your lifetime. Horses have brought so much – meat and milk, but also transport, trading and raiding, and connections across a wider landscape: you started to make contact with faraway people you've only heard about in stories. All those things that seemed impossible when you were a child are now part of everyday life, as if it's always been this way. You would think nothing of travelling thirty, forty miles in a day to see your cousins. You'd think nothing of travelling that far to raid other camps and steal their copper and their animals.

Your children have grown up riding horses as if this was an entirely natural state of affairs. And now, just a couple of decades after that first, exhilarating ride, it isn't just your tribe that are horse people. The idea caught on like wildfire. You've given horses as gifts to three tribal leaders, securing their friendship and allegiance. As young women have left the tribe to marry into others, they've taken their horses with them. Across the steppe, the bond between people and horses has rippled out and taken hold. More wild horses have been caught and broken in, and every year new foals are born to tamed mares.

First steeds

No one really knows exactly how or why horses were first domesticated – but archaeology provides us with clues. The geography of horse domestication maps on to the steppe – where these grazing animals could continue to thrive even while much of Europe had become blanketed in forest. Up on the Great Eurasian Steppe, humans and horses had shared the landscape for tens of thousands of years. Around five and half thousand years ago, the relationship – which had previously been that of hunter and hunted – was set to change, and the fate of *Equus caballus* and the trajectory of human history would become deeply intertwined.

The 'kitchen waste' of archaeological sites is hugely informative. We can find out from this waste exactly what people were eating. In

Mesolithic and Neolithic sites in Europe, horse bones typically make up just a small percentage of animal bones. But on the steppe, such archaeological sites contain plenty of horse bones – around 40 per cent. Humans living there depended on these animals – and would have been very familiar with them – long before they ever caught and tamed them.

Horse domestication comes much later than cattle domestication. And by 7,000 years ago, cattle herding had reached the Pontic-Caspian Steppe. Foragers around the Dnieper River, running down to the northern coast of the Black Sea, were coming into contact with the farmers, who were spreading north and east, bringing their cattle, and also their pigs, sheep and goats with them.

Still, cattle herders could have carried on hunting wild horses rather than domesticating them. The anthropologist David Anthony has suggested that an icy climate could have been the driver. Cattle and sheep are fairly useless at digging through snow to get at forage, especially if that snow is crusted with ice. And they don't tend to break ice to get to water, either. But horses use their hooves to do all of this. They are well-adapted creatures of the cold grasslands. Anthony suggests that a climatic downturn between 6,200 and 5,800 years ago could have seen herds of cattle struggling to make it through harsh winters – and perhaps this is what drove the cattle herders to catch the equine denizens of the steppe. Alternatively, it's possible the domestication of horses emerged more naturally out of a horse-hunting culture. Maybe people who had been hunting horses for centuries, millennia – who *understood* them – started to catch and ride horses, in order to hunt other wild horses. But even that sounds too thoughtful, too strategic. Surely the first people who jumped up on to the backs of wild horses were teenagers, daring each other to do this unthinkable, foolish, brave thing.

In the early Neolithic, the people of north Kazakhstan were still mainly foragers, living in temporary camps. They hunted a variety of wild animals, from horses and short-horned bison to saiga antelope and red deer. But in the 1980s excavations at a site called Botai appeared to reveal a shift, happening around 5,700 years ago, to more specialised horse-hunting. At the same time, the people of the Botai culture, as it has become known, had also adopted a semi-sedentary lifestyle – they certainly don't seem to have been nomads, following herds of wild horses. They were much more settled than that.

The vast majority of animal bones from Botai and similar sites dating to the fourth millennium BCE are all from horse. It's clear that the Botai were eating *a lot* of horse meat. The evidence suggested that Botai people were not only able to trap entire herds of horses, they could also transport carcasses back home. This is a crucial piece of the puzzle: those horses were not killed and butchered *in situ*, like the horses at Wally's Beach – they were brought back to the settlement. Archaeologists argued that the Botai *must* have been riding horses to hunt, and using horses as transport. But as more evidence emerged, the interpretation of the finds from Botai and related sites started to change. Amongst the archaeological remains at Botai sites, there are few spearheads, but plenty of what appear to be leather-working implements – bone tools showing characteristic patterns of microscopic wear. These clues suggest that the Botai were keeping horses – and riding them – not just hunting them. Archaeologists went further, probing the evidence, to test that suggestion.

Although there are only subtle differences in the shapes of bones between different equine species, and between wild and domestic horses, the metapodial or cannon bone, from the lower leg, has been shown to be a particularly informative part of the skeleton. So the archaeologists compared the shape of horse metapodials from Botai sites with those from other localities and periods. They found the Botai bones to be quite slender – and similar to those found at later sites, containing what were definitely domestic horses. They were also similar in their slenderness to metapodials from modern Mongolian horses.

Then they turned their attention to the teeth of the Botai horses, and found something quite extraordinary. They discovered a band of wear on the front edge of one premolar – the enamel of the tooth had worn right through, down to the dentine. If you've looked a horse in the mouth (it's OK if it's not a gift horse), you'll have noticed that there's a gap between its front teeth and its back teeth – it's known as the 'bars' of the mouth, or the diastema. The only thing that could have caused this pattern of wear on the tooth was something which had been regularly placed in the bars of the mouth of that Botai horse – it had clearly worn a bit. Two other teeth showed more subtle signs of possible bit wear. The tooth with the very clear signs of wear was radiocarbon dated to 4,700 years ago. There were also bony growths on the surface of four other mandibles, in that gap between the teeth, just where the bit would sit in the horse's mouth.

Finally, the archaeologists turned their attention to the pottery from Botai sites. They analysed the residues on the inner surface of sherds of cooking pots, and found evidence for not only horse fat but, in particular, lipids from horse milk. While hunters of wild horses would have undoubtedly tasted horse milk occasionally, when a lactating mare was killed, the milk on those cooking pots points to a more regular consumption. Far away from the centre of sheep, goat and cattle domestication and dairying in the Fertile Crescent, the people of the Eurasian Steppe independently came up with their own form of dairy farming. And it was a way of life, an economy, focused on horse meat and milk, that would continue for a very long time in Kazakhstan – right up to the present day. The herdsmen of the Altai Mountains are the inheritors of that ancient way of life, and fermented mares' milk, in the form of kumis, is still a popular drink on the Eurasian Steppe.

This hat-trick of three separate strands of evidence – the leg bones, clear signs of bit wear, and use of mare's milk – all point to the same thing. The Botai of ancient Kazakhstan were harnessing, milking and keeping domesticated horses by the fourth millennium BCE. But it doesn't mark the *start* of something. It's what archaeologists call a *terminus ante quem* – what it tells us is, by this point in time, domestication had happened.

The bit wear shows that the Botai horses were harnessed – bridles could have been used for driving them, but perhaps more likely, riding them. Beyond this specific evidence for keeping domesticated horses, the Botai culture itself goes back to 5,500 years ago. And it's likely that horse-riding started even earlier than this. Burials in the Pontic-Caspian Steppe dating back to 6,500 years ago contain the skeletal remains of horses alongside bones of cattle and sheep. There's clearly a symbolic association between these animals. It's prompted archaeologists to suggest that horses may have been ridden this early, to herd the other animals.

Other clues turn up in the Danube Delta, in modern Romania and the Ukraine, with the appearance of horse-head stone maces and burial mounds, or kurgans – characteristic of steppe cultures – dating to 6,200 years ago. It strongly suggests that horse-riders from the steppe were moving south. Inside the kurgans, the dead were buried with necklaces of shell and tooth beads, as well as axes, twisted neck-rings and spiral bracelets made from a new material – that they'd acquired by trading with the people of the old European towns around the Danube – copper. They'd bought into the Aeneolithic, the Copper Age, and this shiny, malleable metal had become an emblem of prestige. This early expansion of steppe people may have carried something of its own, along with the horses: they may have been speaking a Proto-Indo-European language, a language which would evolve into Anatolian as they moved even further south.

So it seems likely that taming and riding horses could have started a thousand years before the Botai culture emerged – perhaps as early as the fifth millennium BCE. By the fourth millennium BCE – 5,500 to 5,000 years ago – horse bones were already becoming more frequent around the Caucasus, the mountainous region that stretches between the Black and Caspian Seas, south of the steppes. The same thing was happening in the Danube Delta, to the west of the Black Sea. By 5,000 years ago, the frequency of horse bones at some sites in central Germany had risen to account for 20 per cent of all animal bone. The connection is clear: horse-riding and domesticated horses were spreading – fast. Horses and horse-riding spread south of the Caucasus, too. After 5,300 years ago, horses are found more frequently in Mesopotamia – just as the Sumerian civilisation began to blossom.

Riding horses wouldn't just have helped with horse husbandry – it would make herding other animals much more efficient as well. One person on foot, with a good dog to help, could herd 200 sheep. On horseback, with a dog, you could control 500 – and cover a much larger area. Expanding territories would surely have brought pastoralists into conflict with each other. Building alliances and gift-giving would have become important. The proliferation of copper and gold jewellery in the archaeological record suggests that people were seeking status and displaying wealth in a way they simply hadn't before. But it all came at a cost: polished stone maces – some in the form of horses' heads – also begin to turn up at this time. Riding and warfare seem to have been intimately linked, even at this early stage. Formal cavalry might not have emerged until the Iron Age, around 3,000 years ago, but mounted raids – stealing animals from other tribes – and the internecine strife that went with them probably go back to the very dawn of horseback riding.

Towards the end of the fourth millennium BCE, the herders of the steppe were becoming more mobile again. A climatic improvement in the early centuries of that millennium was followed by a downturn. Large herds now needed to roam more widely in order to take in sufficient pasture. This seems to have stimulated the emergence of a new way of life, and a new culture. The herders couldn't afford to be semi-settled any more, as they had been at Botai; they needed to move with their herds and flocks. The solution: wagons. These wheeled vehicles first appeared on the steppe around 5,000 years ago. This sounds very precise. How on earth could archaeologists make such a claim, about vehicles which leave so little trace on the ground? Wheel ruts don't tend to last for millennia (and where they are found, they're impossible to tell apart from sled-ruts).

The answer lies inside the graves of these steppe people. They were still making kurgans, and under those mounds of earth, they buried their elite – mostly men – *with their wagons*. These extraordinary burials, with bodies and dismantled wagons placed in pits, appear across the Pontic-Caspian Steppe, dating to between 3000 and 2200 BCE. The burial rite gives this new culture its name, though it's a name those people never knew – 'Yamnaya', after the Russian for 'pit-grave'.

The wheel itself may not have been invented on the steppe. It's thought that the idea of wheeled vehicles spread there, either from the west, from Europe, or from the south, from Mesopotamia. The earliest known image of a wheeled vehicle comes from a site in Poland, and dates to around 3500 BCE, while a clay model of a wagon from Turkey dates to about 3400 BCE. With covered, ox-drawn wagons as their mobile homes, the herders could migrate around the landscape, following huge flocks and herds. And they were still riding horses, of course. Archaeologists suggest that the cycle of the year would have taken the herders out on to the open steppe in spring and summer, while in winter they'd set up camp in river valleys. Crucially, those valleys would have contained trees – providing wood for fuel, and for mending the wagons. Although this horse-riding, wagon-camping, kurgan-building culture stretched right across the Pontic-Caspian Steppe, there were regional differences in livestock and the types of plant being eaten. In the east, beyond the River Don, people were mainly tending sheep and goats, with just a few cattle and horses – which provided essential mobility. Along with mutton and goat, they were eating foraged tubers and the seeds of goosefoot – a plant very closely related to quinoa. In the western steppe, people were more settled – and they were herding cattle and pigs, and growing some cereals.

But – like the earlier horse-riding nomads in the fifth millennium BCE – the Yamnaya didn't stay on the steppe. By around 3000 BCE, they had begun a westward expansion, into the Lower Danube Valley, and on to the Great Hungarian Plain. The herders of the steppe were expanding eastwards as well, making contact with the early farmers of China. Crops and animals that had been domesticated in the west spread east. The idea of copper metallurgy may also have travelled to China from the west. After the Yamnaya, there seem to have been successive waves of steppe people expanding both east and west. Over five thousand years, this scenario played out again and again, with the last wave recorded in history – as the thirteenth-century Mongol invasion.

The prehistoric expansions of steppe nomads appear to have had very different impacts on existing societies in the east and west. In China, the nomads seem to have merged with settled societies, but in the west they encroached on lands occupied by other nomadic pastoralists – and they caused a knock-on effect, pushing those nomads even further west.

The Yamnaya expansion into Europe had a profound cultural impact, one that is still echoing today. Geneticists and comparative anatomists use patterns of similarity and difference between modern organisms, and ancient ones too, when they're available, to construct phylogenies – family trees which represent evolutionary history. Linguists can do much the same thing with languages, using comparative grammar and vocabulary. Many ancient and modern languages, from English to Urdu, Sanskrit to ancient Greek, all group together, in the Indo-European language family. And linguists have traced the evolution of sounds back and back, until we have the closest thing to an original Indo-European language – in the form of about fifteen hundred distinct sounds. It's very hard to test whether they really have found the traces of an ancient language, but archaeological discoveries have since revealed previously unknown words in Hittite and Mycenaean Greek – which were correctly predicted by the methods of the historical linguists, giving us some grounds to trust their reconstructions.

The fragments of the Proto-Indo-European language contain words for otter, wolf and red deer as well as for bee and honey, cattle, sheep, pig, dog and horse. In other words, it's a language root that clearly emerged after the beginning of the Neolithic – its speakers had words for domesticated animals. Still, it's not clear that the word for 'horse' refers to a domesticated horse. But there are other clues. The reconstructed Proto-Indo-European also contains words for wheel, axle and wagon. It seems that the Yamnaya people, the horse-riding, wagon-driving nomads of the steppe, were speaking a language which would form the basis for all the Indo-European languages that we continue to speak, across Europe and western and south Asia, today. How wonderful to think we're still using words that contain a faint echo of that ancient culture on the steppes.

Close cousins and alternate histories

Unravelling the origins of domestic horses has proved – this is now a familiar theme – very difficult indeed. Just as with wolves and early dogs, aurochsen and early cattle, it's difficult to discern any differences between the bones of wild and domestic horses. Those metapodials from the Botai sites were only subtly different to wild horse bones. In fact, there are very few differences in the skeletons of any species belonging to the genus *Equus* much more generally. If you compared the skeleton of a zebra with the skeleton of an ass, you'd be hard-pushed to tell them apart. Once again, this is where genetics has ridden to the rescue. And before looking at the origin of domestic horses, we need to make sure that we really understand the differences between the various equine species that exist today. Some have recently turned out to be even less 'different' than we'd thought.

It seems that we've been taxonomically over-zealous in dividing up *Equus* in the past. Genetic analyses have prompted suggestions that some apparently discrete populations, traditionally seen as separate species, are actually much more closely related. For instance, the plains zebra and the extinct quagga are traditionally labelled as separate species, based largely on how they look. Modern genetics says: they are one and the same species. Similarly, the extinct 'stilt-legged horses' of America are, genetically, caballines – close cousins of modern domestic horses. But, as the family tree of horses collapses in on itself, with fewer branches and closer genetic relationships than previously suspected, one key part of the family tree is uncontested – and that's the very close relationship between domestic horses, *Equus caballus* – together with their wild ancestors and relatives, *Equus Ferus* – and the surviving, truly wild horses of the central Asian steppe: *Equus przewalskii* – Przewalski's horse. These small but robust horses have a sandy or reddish coat, pale muzzles and bellies, a bristly brown mane, and a stripe along the back.

Genetic analyses have made it possible to reconstruct the family history of the caballines, or true horses, and pin some dates on it. The wild ancestors of domesticated horses emerged as a separate lineage some 45,000 years ago, when they diverged away from the ancestors of Przewalski's horse – long before domestication. Despite that divergence, however, a small amount of interbreeding continued – the reciprocal gene flow shows up quite clearly in genomes today. Most of that interbreeding happened long ago, before the last glacial maximum, twenty thousand years ago. After the Ice Age, there was still some genetic input from Przewalski's horses into the ancestors of domestic horses, and it continued even after domestication. Later still, right at the start of the twentieth century, there's evidence for gene flow in the other direction – from modern horses into Przewalski's horses. This last injection of domestic horse genes into Przewalski's horse was precisely at the time when Przewalski's horses first started to be kept and bred in captivity.

The ability of these two horse populations to interbreed really is extraordinary. They're distinct enough – morphologically and genetically – to be considered separate species. And they have different numbers of chromosomes – often considered to be a complete barrier to interbreeding. While domestic horses have sixty-four chromosomes (thirty-two pairs), Przewalski's horse has sixty-six (thirty-three pairs). When a mammalian egg or sperm is made, it ends up with half the usual genetic complement found in other cells of the body. At fertilisation, the genetic material from the egg is combined with that from the sperm to create a full set once again. Each chromosome from the egg must pair up with its opposite number, from the sperm, before the fertilised egg can start to divide and make an embryo. If a domestic horse and a Przewalski's horse mate – the resulting fertilised egg would have one set of thirty-two chromosomes and one set of thirty-three. But somehow (and even geneticists are astounded by this), the chromosomes manage to pair up – if they didn't, there'd be no viable offspring. And the traces of interbreeding in the genomes of modern domestic horses and Przewalski's horses shows that, not only were such offspring viable, they were fertile – they could reproduce.

Of course, hybrids between equine species are well known. A hinny is a cross between a male horse and a female donkey. A mule is a cross the other way – between a male donkey and a female horse. And although hinnies and mules are usually sterile, they can occasionally reproduce successfully. That's again quite remarkable considering that donkeys have thirty-one pairs of chromosomes and horses have thirty-two. But the genomes of different equine species contain evidence of an even more astounding feat – interbreeding and gene flow between the Somali wild ass, with thirty-one chromosome pairs, and Grévy's zebra, with twenty-three. Such findings challenge our stereotyped views about how biology should work. The species boundary is turning out to be much more permeable than we'd anticipated, before genomics came along. Even differences in chromosome number don't seem to be quite the barrier to successful reproduction that we'd previously imagined.

As well as tackling questions of interbreeding, genetics provides an insight into fluctuations in the size of ancient populations over time. Both the ancestors of domestic horses, also known as *Equus ferus*, and Przewalski's horse suffered a population crash in the late Pleistocene and early Holocene, around 10,000 to 20,000 years ago. Populations continued to dwindle – up until the point of domestication, some 5,000 years ago. Then, for *Equus caballus*, the domesticated horse, the future started to look decidedly rosy. But while this population of horses grew and grew, and spread around the world, their wild cousins would become endangered.

The close wild relative of domestic horses, *Equus ferus*, also known as the Tarpan – with a characteristic sandy-grey-coloured coat, pale belly, black legs and a short mane – finally went extinct in 1909. The Przewalski's horse was also spiralling towards extinction. These rare, shy horses were spotted by a Russian explorer and geographer, Nikolai Mikhailovich Przewalski, who was making his way across the central Asian steppe in 1879. By this time, the range of these wild horses had contracted, and there were only small herds roaming the steppe in what is now Mongolia and Inner Mongolia. When Przewalski was preparing to leave Mongolia, he was presented with the hide and skull of one of the horses, which had been shot, and which he duly took back to St Petersburg. These remains were studied by the zoologist I. S. Poliakov, who published his description of this unusual horse in 1881. Poliakov determined that the remains of the beast from Mongolia were sufficiently different from domesticated stock to be considered new to science, and he gave the wild horses of Mongolia a new species name, honouring its discoverer. The horses immediately became collectable, and expeditions set out to Mongolia to capture specimens for zoos – further depleting the wild population. The last Przewalski's horse to be caught was a mare, named Orlitza, captured as a foal. The horses were getting rarer and rarer in the wild. Being recognised as a new species was in some ways their downfall. The expeditions that fed zoological collections inevitably killed some animals, and dispersed others.

In 1969, the last sighting of a wild Przewalski's horse was reported, from the Dzungarian Gobi in south-western Mongolia. Extinct in the wild, a few Przewalski's horses survived in zoos, long enough to breed. In the 1980s and 1990s, attempts were made to reintroduce these horses to the wild, using horses bred from a stock of just fourteen individuals – including Orlitza. And the attempts were successful. Between the 1960s and 1996, Przewalski's horse was considered 'extinct in the wild', but by 2008 it was back in the game – though in such small numbers it was designated 'critically endangered'. The numbers crept up – by 2011 it was considered to be merely 'endangered' – which meant that a population of more than fifty mature animals was living wild.

It's now estimated that there are a few hundred Przewalski's horses living outside of captivity. These small numbers mean that the population remains very vulnerable to adversity – to disease and severe winters – but they have some help. In the Kalamaili Nature Reserve in Xinjiang, China, where Przewalski's horses were released into the wild in 2001, the horses are rounded up into a corral each winter to give them extra food and to protect them from competition with domestic horses. By 2014, there were 124 individuals in just this one group of rewilded Przewalski's horses, which has been described as the most successful of the reintroduction efforts in China.

The population in captivity is looking healthy too – with around 1,800 horses in zoos worldwide, and growing. The reintroductions to the wild have mainly taken place in China and Mongolia, around the area where the horses were last seen living wild before going extinct. But some Przewalski's horses have also been released into nature reserves and national parks – in Uzbekistan, Ukraine, Hungary and France.

The story of these wild horses provides us with an alternate history for their domesticated cousins, *Equus caballus*. What would have happened had these horses remained wild? There's no doubt that the course of human history would have been altered, but the fate of horses would have been significantly affected as well. Wild horses formed an important source of meat for our Palaeolithic hunter-gatherer forebears, but they are surely very likely to have been hunted to extinction, or close to it, had they not proved useful to our ancestors in other ways – carrying riders across the vast steppes and knights into battle; pulling chariots, wagons and cannons; becoming emblems of status and prestige in human society. The reintroductions of Przewalski's horses into the wild seem to be a success story – a triumph for rewilding – but the global population of these horses, out in the wild and in captivity, numbers just a few thousand at most. In comparison, there are around 60 million domestic horses on the planet. There are concerns about diminishing genetic diversity amongst them, and about loss of breeds, but *Equus caballus* is far from being an endangered species.

Leopard spots and horses' faces

Perhaps it seems that we've already nailed the origin of this particular domesticate. Yet while the earliest archaeological evidence for domesticated horses undoubtedly comes from the Pontic-Caspian Steppe, this doesn't mean that all our horses today must have come from a single origin. There could have been later, independent centres of domestication. After all, horses ranged widely across Eurasia – there were plenty of other places where horses and humans had been in contact for millennia. In a more diffuse, multiregional model, separate herds would later have merged into a single, diverse population of domesticated horses, which continued to reflect regional differences and local origins. Just as with dogs, the apparent diversity of modern horses could be taken to suggest that multiple origins were most likely. In the past, similarities between some domesticated breeds and local, wild ponies were taken to support such a model. Studies of morphological characteristics – the shape and size of bones – suggest a strong similarity between the Exmoor pony, the Pottok pony of the Basque country, and the extinct Tarpan. Some have argued that the beautiful, semi-wild horses of the Camargue represent the direct descendants of the ancient, truly wild Solutré horses which featured in Ice Age cave paintings. But the genes say something different, and more interesting.

In 2001, a piece of research was published, based on analysis of a particular stretch of mitochondrial DNA in samples taken from thirty-seven individual horses – and this bit of DNA was very variable indeed. But did this diversity represent lineages which had diverged from each other before or after domestication? If before, this would suggest multiple origins for modern horses. If after, that would support a single origin. In order to answer this question, the geneticists looked at the mitochondrial DNA of a donkey – which was 16 per cent different to that of horses. They assumed that donkeys and horses had split apart from each other some time between 2 million years ago (as the fossil record suggested) and 4 million years ago (according to the genetic estimates at the time). This gave them a form of calibration – over 1 million years, you'd expect to see genetic sequences diverging by 4 per cent (given a split 4 million years ago) to 8 per cent (for a 2 million-year-old split). They could then apply this rate to the differences they'd found within modern horse mitochondrial DNA, which amounted to about 2.6 per cent. The calibration suggested that the lineages found in those modern horses must have been diverging for somewhere between 320,000 to 630,000 years. Even the lower estimate places the origin of this genetic diversity *well* before the date of domestication around 6,000 years ago. The geneticists went on to suggest that wild horses had been captured over a huge area, and used for both meat and transport. Then later on, as wild populations began to disappear, domesticated herds became more important and were interbred, to form the genetic basis of modern horses. The researchers contrasted the domestication history of horses with that of dogs, cattle, sheep and goats. Firstly, those other species had been domesticated much earlier (still true), and they all came from a restricted origin, then spread out. The domestication of horses, on the other hand, seemed to have happened again and again, and in many places it represented the dissemination of an idea – a technology – rather than a spread of the animals themselves.

But that's the mare's tale. What about the stallion's? It turns out they have a completely different story to tell. The anthropologist David Anthony describes horses as 'genetically schizophrenic'. Maternally inherited mitochondrial DNA suggests that modern, domestic horses come from a great variety of wild female ancestors. The mitochondrial genetic diversity of horses is prodigious – and very unusual compared with other domesticates. But the paternally-inherited Y chromosome records only a very few wild, male antecessors.

The discrepancy between mitochondrial and Y chromosome data may – to some extent – reflect natural breeding patterns. Both Przewalski's horses and feral horses operate in harems. This seems to be the natural state of things in horse society: polygyny is the norm, with one, dominant stallion presiding over a herd of mares and foals. Young males leave the herd and hang around in bachelor groups for a few years before they attempt to form their own harems – either by stealing mares from other stallions, or fighting to take over an existing harem. And so the genetics of modern horses may reflect social and reproductive patterns which are natural in horse populations.

But actually this isn't enough to explain the stark contrast between the variation seen in mitochondrial lineages, compared with the Y chromosome. This pattern strongly suggests that many more mares were domesticated than stallions. To me, this makes so much sense. Stallions are, by their nature, feisty, independent, even dangerous. It would always have been difficult to find a young, male wild horse that wouldn't go crazy, throw you off, kick you in the head. Mares are naturally more docile. If you're a herder looking to catch and tame a wild horse, you've got a much better chance with a mare. So it's not surprising that, historically, more wild mares than wild stallions were caught and tamed. But whilst mares were easier to tame, at least one stallion would have been needed for successful breeding. That's basic biology.

Looking at the DNA of modern horses, however, there are missing parts to the puzzle. You don't know where and when particular lineages were added to herds of domesticated horses, and you don't know how much past diversity has been lost over time. Ancient DNA – extracted from age-old bones – provides much more depth to the story. Towards the end of the Ice Age, there was a large, genetically connected, population of wild horses ranging from Alaska to the Pyrenees. By 10,000 years ago, the North American horses had disappeared, and the population on the Eurasian Steppe had become separated from the one in Iberia. The latest genetic studies also reveal that Y chromosome diversity has been lost over time – giving us the mistaken impression that only a few stallions were ever domesticated.

While ancient and modern DNA is consistent with horse domestication *starting* in the western Eurasian Steppe, in the Copper Age, it also shows maternally inherited, mitochondrial DNA – from wild mares – entering domestic herds again and again, as domestic horses spread across Europe and Asia. More wild mares were caught and domesticated in the Iron Age, and again in the Middle Ages, adding their wild genes to the already domesticated herds.

A few maternal lineages found in ancient, pre-domestic Iberian horses – walled off from the rest of Europe by the Pyrenees – found their way into the domestic stock, and are still there in some Iberian breeds today, such as Marismeño, Lusitano and Caballo de Carro. As it was the Spanish who reintroduced the horse to South America, it's not surprising that this ancient Iberian signature also turns up in South American breeds, such as the Argentinean Creole and Puerto Rican Paso Fino. But it's also found in some French and Arabian horses, probably reflecting ancient trade between Iberia and France, and the close connections between Spain and North Africa. In China, most of the mitochondrial DNA lineages point to domestic horses spreading to East Asia from further west – but then with a few lineages being added from local, wild populations.

Rather than multiple, independent centres of domestication, the picture that's now emerging is one of domesticated horses spreading from their original homeland in the steppes – but with plenty of wild mares being added to the existing domestic herds along the way, and through history. So it wasn't just a spread of an idea and a new technology after all – it was a spread of horses, too.

Like other domestic species, the story doesn't stop there. Selective breeding has promoted certain traits while suppressing others. Just as with dogs, cattle and chickens, strong artificial selection, brought about by strict breeding regimes, has operated over the last two centuries to create the range of modern breeds we know today. But selection was happening way back in the past, too. Small horses, built for speed and agility, were favoured for pulling light chariots – a Bronze Age invention – while the Iron Age Scythians bred larger horses, selecting for endurance in some, and speed in others. Medium-weight horses were pressed into battle service pulling wagons and, later, artillery. By the Middle Ages, draught horses had become massive, weighing up to 2,000 pounds.

Some traits which appear in our modern breeds were already there in pre-domestic horses. The horses running across the walls at Lascaux are brown and black – these could easily have been naturalistic coat colours. The spottiness of the horses at Pech Merle has been suggested to be rather imaginative, perhaps symbolic, even psychedelic – fitting in with the abstract spot patterns that appear around the horses. But on the other hand, the dappling of the Pech Merle horses looks very much like the 'leopard' pattern of coat seen in some modern breeds, such as Knabstrupper, Appaloosa and Noriker. The genetic basis of this 'leopard' pattern is well known – it's down to a particular variant or allele of the LP gene on horse chromosome 1. Geneticists screened DNA from a range of thirty-one pre-domestic, ancient horses from Europe and Asia to see if they could pick up this variant. None of the Asian horses had the LP allele, but it was there in four out of the ten western European horses they looked at. There's definitely some artistic licence at work in Pech Merle – the horses have particularly tiny heads and spindly legs – but the dappled pattern could easily represent the real-life appearance of Ice Age horses, copied straight from nature. It also seems to have been a characteristic which was particularly favoured by some early horse breeders – the LP gene turned up in six out of ten horses from a Bronze Age site in western Turkey, for instance.

Up in northern Siberia, it's been suggested that Yakutian horses might have interbred with local, wild horses – imbuing them with key physiological and anatomical characteristics which allow them to survive in subarctic conditions. These horses are compact, with short legs, and they're also incredibly hairy. But in this case, genetic studies of ancient and modern Yakutian horses have revealed no particular connection between them. The modern horses of Yakutia seem to have been introduced in the thirteenth century CE – and they've adapted extraordinarily quickly to a cold environment. Such rapid changes in genes related to hair growth, metabolism and blood-vessel constriction (to lessen heat loss at the surface of the body) must have been crucial to survival. Amongst domestic horses more generally, other genes that show evidence of having been positively selected for in the past appear to relate to changes in the skeleton, circulatory system, brain – and behaviour.

There are some fascinating elements of behaviour that horse owners may have known about, or at least suspected, for a very long time – and which scientific studies are just beginning to elucidate. Evidence suggests that cats and dogs are able to understand human emotions, expressed both physically and vocally. Dogs really do seem to understand what a happy human face looks like. It's known that horses themselves make facial expressions, and that they can recognise emotions in another horse's face. In a recent study, horses were shown pictures of men making angry, frowning faces and happy faces. The heart rate of the horses tended to increase more when they were looking at an angry face, compared with a smiling one. If this means that horses really do read human emotions, there are a few explanations for this ability. It could be that horses, having been able to interpret expressions on other horses' faces for a very long time, started to be able to do the same with humans after they were domesticated. Or it could be something that horses learn to do during their lifetimes, associating other cues indicating angriness with an angry human face, for instance. This could still stem from an inbuilt predisposition to inferring emotion from physical appearances, inherited from their wild ancestors.

Another recent, carefully constructed study showed that not only are horses able to interpret our behaviour, they attempt to influence it as well: some gestures made by horses really do seem to be an intentional form of communication. Horses were seen to stretch their necks, pointing their heads towards a bucket full of food they liked, but could not reach. They'd look at the human experimenter, then 'point' at the bucket, then look at the experimenter again. If the experimenter walked away, they'd stop. When the experimenter walked towards them, they'd alternate their gaze more frequently. The horses also used nods and shakes of the head to get attention. It suggests that horses not only wish to communicate, they recognise humans as capable of receiving communication. It's unlikely that horses have evolved to be able to do this over just a few thousand years of being domesticated, but it's also unlikely that it's innate. Instead, horses are probably predisposed to learn this sort of behaviour, through interacting with other horses – and now humans as well – in their social environment. So whilst the behaviour isn't innate, the tendency to develop it is. Being naturally sociable animals, much like dogs, meant that horses were well suited to teaming up with another, sociable animal. They've made good allies, ever since the Copper Age, when the horse-hunters of the Pontic-Caspian Steppe became horse-riders. They became fellow-travellers – but it wasn't only humans that they were carrying. The beginning of the diaspora of our next domesticated species started with the saddlebags of travellers along what would become the western end of the Silk Road. Stuffed into those saddlebags were fruits for the journey – apples.

9

APPLES

Malus domestica

Wassail! wassail! all over the town,
Our toast it is white and our ale it is brown;
Our bowl it is made of the white maple tree;
With the wassailing bowl, we'll drink to thee.

Gloucestershire wassail

Wassailing

It's late January, and a cold evening in north Somerset. A small crowd of people gathers in an orchard. Bare twigs stretch up into the night sky. Ice crystals are already crunching underfoot. Everyone, young and old, is well-coated, and muffled up with scarves and woolly hats. Breath comes out steaming in the frosty air. The kids have got instruments. It's hard to call them 'musical' – they're anything that will make a noise: maracas; tambourines; tin cans with bottle-caps inside; more bottle-caps strung on to a bit of wire attached to a forked stick, as an improvised rattle. One of the adults has got a trumpet. The crowd starts to move, becoming a snaking procession, making its way under the trees, with much jangling and banging and shaking. It's an almighty racket.

We're waking up the cider trees and scaring away evil spirits, to ensure a good harvest when autumn arrives. The procession comes to a halt and a man clears his throat to begin the wassailing song. People bursting into song in public has always made me feel deeply uncomfortable. It's *showing off.* It's like having to watch other people's children perform a play they've just made up that afternoon. You can't escape, and it's impolite to laugh. You have to sit there, smile encouragingly rather than grimace, and then congratulate them afterwards, with not a note of irony. But here – in this orchard – my icy cynicism melts a little. The man has a beautiful, mellifluous, age-old voice, and he throws himself into this performance wholeheartedly. I feel as though we're slipping through time, re-enacting, re-voicing the echo of something that has been happening for centuries.

Then we all troop back into the house. We shed the woollens and the coats and the past. We start chatting to friends, and we're free from the spell, back in the now. But we'll still grab a cup of mulled cider and drink each other's health, and that's another ancient echo. This tradition of wassailing goes back to at least Medieval times – but its roots probably go back deeper into antiquity. It's an unashamedly pagan ritual, designed to propitiate the tree spirits and secure a good harvest. The first recorded mention of a wassail is from 1585, from Kent, where young men were rewarded for wassailing in the orchards. In the seventeenth century, the writer and antiquary John Aubrey recorded a West Country custom where men went into the orchard with a wassail bowl and went 'about the trees to bless them'. In the eighteenth century, wassailing rhymes and songs proliferated. The nineteenth century saw a steep decline. In the twentieth century, revivals of the old rite were variably successful. It seemed to cling on most tenaciously in the Welsh and English counties around the Severn. My friends' wassail in their orchard was a contemporary echo of a long tradition, albeit a revived one.

'Wassail' comes from Old Norse – '*ves heil*', meaning 'be you healthy'. As we retreat inside and drink each other's health in warm, spiced cider, we're marking the beginning of a new year, and hoping that it will turn out well for our friends, and for the apple harvest.

Apples are so quintessentially English. The wassail celebrates and underlines our primeval connection with these trees and their fruit. But apples – like all the other domesticated species in this book – don't come from this little island in north-west Europe. The original homeland of apples was over three and a half thousand miles away.

On the flanks of the Heavenly Mountains

We've been here before, or at least very close. The region of Dzungaria is named after an ancient Mongolian kingdom, and most of it is now encapsulated by the Chinese province of Xinjiang – sandwiched between Kazakhstan to the west and Mongolia to the east. But the eastern tail of old Dzungaria still lies in Mongolia, and this is where the last Przewalski's horse was spotted in 1969, before the species disappeared. To the south, Dzungaria is bounded by the Tian Shan Mountains. The mountain range continues to the west, expanding into an altitudinous wedge of land which forms modern-day Kyrgyzstan, separating Kazakhstan in the north from the south-western projection of Xinjiang Province in China.

A fertile oasis amongst steppe and desert. 'Tian Shan' means 'heavenly mountains', and they seem to live up to their name. Botanist Barrie Juniper described their beauty: 'With its jagged, glistening, snow-covered peaks, forest-clad slopes, and high, sheltered pastures bejewelled in spring with flowering bulbs and fruit blossoms, and in the autumn with a cornucopia of fruit, the Tian Shan is the apotheosis of a favoured, ancient mountain kingdom.'

In 1790, a German pharmacist-botanist called Johann Sievers joined a Russian expedition to southern Siberia and China, in search of a particular species of medicinal rhubarb. But he wasn't so obsessed by his quest for rhubarb that he ignored the other plants he found along the way. On the flanks of the Tian Shan, in what is now south-eastern Kazakhstan, Sievers found forests of huge apple trees, laden with unusually large and colourful fruit – some green or yellow, others red and purple. These weren't mixed deciduous forests with an occasional apple tree: apples were the predominant species. And they weren't the dwarfed and pruned apples of our modern orchards – these apple trees grew up to 60 feet high. Sievers himself died very soon after returning from the expedition, at the age of just thirty-three, before he'd had a chance to describe his discovery – but he was later immortalised in the botanical name of the apple that he'd found in the heavenly mountains of central Asia: *Malus sieversii*.

In the early nineteenth century, botanists and apple growers were trying to make sense of the twisted branches of the genus *Malus*. The large-fruited trees of the forests around the Tian Shan seemed to have been largely forgotten about. Instead, the prevailing idea was that cultivated apples had been domesticated from European wild apples – including *Malus sylvestris*, the 'woodland apple'; *Malus dasyphylla*, the south-east European 'Paradise apple'; and *Malus praecox*, the 'primitive apple'.

In 1929, thirteen years after his Persian expedition to track down the origins of wheat, Nikolai Vavilov, widely regarded as the world's greatest plant-hunter, set off in the footsteps of Sievers. He trekked to south-east Kazakhstan – which by then had been swallowed up by the expanding Russian Empire. There, around the city of Almaty, in the foothills of the Tian Shan, he explored the wild apple forests. Today, Almaty is the largest city in Kazakhstan, home to nearly 2 million people, but its name enshrines its ancient connection with apples. The Russian version of the name, *Alma-ata*, means 'Father of Apples'. The first reference to the city in literature, from the thirteenth century, calls it Almatau, which is thought to mean 'Apple Mountain'.

Vavilov wrote, 'All around the city, one could see a vast expanse of wild apples covering the foothills and forming forests.' He was struck by the similarity of the fruit of some of the wild-apple trees there to cultivated varieties. These wild apples were not like the small and sour-tasting European wild crabapples, but plump fruit, bursting with flavour. 'Some trees are so good in the quality and size of their fruits that they could be transplanted straight into a garden,' he enthused. This was quite astonishing – especially given how different

domesticates usually are from their wild precursors. Just think about
the difference between maize and teosinte, or even domesticated and
wild wheat. Identifying a wild progenitor usually takes a fair bit of
detective work – but not so with the apple. It seemed blindingly
obvious that this wild central Asian species was extremely closely
related to, and shared an ancestry with, the tamed fruit trees of
orchards. Vavilov was sure that the region around Almaty must be the
geographic birthplace of this fruit – the centre of domestication. 'I
could see with my own eyes,' he wrote, 'that this beautiful place was
the origin of the cultivated apple.'

And yet, towards the end of the twentieth century, some botanists
were still focusing on the European crabapple, *Malus sylvestris*, as the
likely progenitor for domesticated apples. Others weren't so sure. In
1993, horticulturalist Phil Forsline, from the US Department of
Agriculture, returned to the forests of south-eastern Kazakhstan.
Joining forces with local scientists, he embarked on a botanical
survey – which included tasting fruits, finding flavours which varied
from nutty to aniseed-like, sour to sweet. He was also keen to collect
seeds from as many varieties as possible, to create an archive of
'germplasm' which could potentially be used to improve crops in the
future. In the end, he and his team returned to the States with more
than 18,000 apple seeds.

Like Vavilov and Sievers before him, Forsline was struck by how
similar some of the wild apples were to domesticated varieties. But
there was another reason to believe that this region around Almaty
represented the original homeland of apples, and that was the daz-
zling variety of the apple trees growing there. Vavilov had realised that
diversity could provide an important clue to geographic origins. As
we've seen, species do tend to be most diverse close to their origin,
where they've had the longest time to accumulate differences. And
large-fruited apple trees seem to have been growing and evolving in
the forests of the Tian Shan for at least 3 million years.

Malus sieversii is a strange fruit tree in many ways. Other species
of wild apple, also known collectively as crabapples, tend to have
small, sour fruit. The origin of the name 'crabapple' is debated – its
Scottish form, scrabbe, suggests that it might have come from a Norse

word, simply meaning 'wild apple', but 'crab' may also have meant 'sour'. Crabapples tend to occur on their own, or in small groups. None form dense forests like *Malus sieversii* on the Tian Shan. Another bizarre feature of this species is its prodigious variation – in the size of individual trees, colour of flowers, and shape, size and flavour of fruit. A key to that diversity may be the depth of time that this species has had to develop there, in the Kazakh forests, but it also tends to rejoice in variety in a way that other *Malus* species just don't. Crabapples are extremely conservative in comparison.

The large-fruited central Asian wild apple seems to have evolved from earlier, small-fruited predecessors, which may have spread across Asia before the Tian Shan Mountains began to thrust up into the sky. And when they did, they formed an island of suitable habitat – a unique physical environment – for an isolated population of apple trees, surrounded by inhospitable deserts. The repeated Ice Ages of the Pleistocene, creating fluctuations in climate across the globe, may have driven plants into fragmented pockets of habitat, over and over again. Perhaps the tendency for wild apples to vary so much – and particularly for offspring to differ widely from their parents – developed as a useful adaptation to environmental variability.

The central Asian wild apple is closely related to the Siberian crab-apple, *Malus baccata*, which has small, red fruits that are eaten by birds, who then do the job of dispersing the seeds – after the pips pass through their guts. It's likely that the ancestor of *Malus sieversii* was also a bird-dispersed apple. But then it diversified. Large fruits strongly suggest that a very different sort of animal – a mammal – has been seduced into helping with seed dispersal. It seems that apples originally grew larger fruit to grab the attention and satisfy the tastebuds and appetite of bears. (Of course, that formulation is a shortcut: the mechanism that has led to the appearance of larger-fruited apples is the mechanism that lies at the heart of evolution – natural selection. Presented with a variety of fruits, bears preferred those that were larger, and the trees bearing them would have had an evolutionary advantage, passing more of their genes on to the next generation of apple trees.) Over time, one lineage of originally small-fruited trees changes into a new species, with large apples that bears just couldn't resist. Small apples would have been less attractive, but also less likely to germinate successfully, if they passed through the guts and emerged relatively unscathed. Apple pips stuck inside apples don't germinate. That seems unnecessarily contrary, but it prevents new apple plants from springing up underneath the parent plant to compete with it. Crunching up larger apples exposes the seeds – an essential step towards germination. If an apple pip escapes being crushed by teeth, it will then pass through the gut intact. When it emerges at the other end, it has a chance of becoming a new tree, perhaps miles away from its parent. Emerging from the back end of a bear, the pip would land in a pile of fertile manure on the forest floor. But being dumped on to the forest floor – even with the bear fertiliser factored in – isn't an ideal position from which to start growing. Luckily, there are other large forest mammals that can help to bury apple seeds: wild boar do a great job of turning and churning soil, increasing the chances of successful germination.

Nevertheless, whilst brown bears (and boars) undoubtedly did an excellent job of spreading apple pips through the forests of central Asia, it would take humans – and their horses – to unleash a diaspora of this fruit across Asia, Europe, and eventually the whole world.

The archaeology of apples

The ancient hunting and gathering nomads of central Asia left scarce, faint traces of themselves behind. Fragments of animal bones from a handful of sites record their presence – so that we know they were mainly hunting horse, ass and aurochs. In the Tian Shan Mountains themselves, there are hints of human presence both before and after the peak of the last Ice Age. As the world warmed up, there was a change in technology. Hunting techniques changed – judging by stone artefacts, dating to around 12,000 years ago, which include tiny 'micro'-blades that must surely have been stuck into the shafts of darts or harpoons to be of any use at all. Then there was that shift from hunting to mobile herding – with the arrival of cattle and the domestication of horses – around 7,000 years ago, close to the start of the Copper Age. By almost 5,000 years ago, in the third millennium

BCE, the Bronze Age had reached the Eurasian Steppe, and recent research has revealed that cereal crops were being grown in eastern Kazakhstan at this time. The grains included wheat and barley from the west, and millet from the east. The people growing these cereals in the mountains were still mobile herders – but they were clearly returning to the same seasonal camps, to sow, harvest and thresh their crops. A chain of Bronze Age sites, from the Yellow River in the east, through the Tian Shan to the Hindu Kush, stand testament to a healthy flow of ideas between east and west way back in prehistory, along what's been named the 'Inner Asian Mountain Corridor'. By the second millennium BCE, pastoralists had moved into the highland valleys of the Tian Shan, bringing their sheep, goats and horses, wheat and barley with them.

The pastoral lifestyle that had taken root in the Tian Shan – and the Altai Mountains to the north-east – may well have been introduced by the Yamnaya. Whether that was by cultural diffusion – a transfer of ideas from one society to another – or by an actual migration of pastoralists to the east, is a hot topic of debate. The first evidence of human settlement at Almaty goes back to the Bronze Age, 4,000 years ago (2000 BCE). Like the Yamnaya, the Bronze Age people of Almaty built kurgans for their dead. Right in the centre of that Inner Asian Mountain Corridor, Almaty quickly became an important stopping-place on the east–west trade routes that connected central China to the Danube – which would later become known as the Silk Road.

Wheat and barley arrived in central Asia from the west, millet from the east. But now it was time for central Asia to offer up its own gift to the rest of the world. People and horses travelling along the proto-Silk Road, which passed through the wild-apple forests, helped to spread apples beyond their homeland, as travellers stuffed the fruit into their saddlebags – or devoured the apples. After all, the fruit of the apple tree had evolved as a means of spreading seeds. It's no accident that apples are delicious – that's part of the way they encourage us to help them spread. Humans – and horses – love eating apples just as much as bears do. And horses could do the job of both bear and boar – splitting the flesh of the apple away from its seeds and depositing those pips in a pile of manure, as well as driving pips down into the sod with their hooves.

And so apples began their diaspora, as freely pollinated, naturally sown seedlings – still essentially wild things, but with new two-legged and four-legged friends to help them on their way. As this fruit spread, it needed a name. Indo-European languages contain two different formulations of a word for apple, one sounding a bit like '*abol*' and the other '*malo*', but it's possible that both of these words come from an original, Proto-Indo-European word – *samlu*. The Bronze Age and Iron Age riders on the Eurasian Steppe may have spoken of apples as *amarna* or *amalna*. It's easy to hear this changing into the ancient Greek *melon*, and the Latin *malum*. But, as it travelled west, the word changed again – with the 'm' becoming 'b'. (This is not such a weird transition as it looks when written down. Make the sound 'mmm', then give me a 'mmmb', then just a 'b' – see? When you perform them, these sounds are all very similar – 'm', 'b', 'p' are all produced by the lips coming together or parting.) The ancient apple word continued splintering into separate languages and dialects, but still the Ukrainian *yabluko*, Polish *jablko* and Russian *jabloko* retain a faint affinity with German *apfel*, Welsh *avall* and Cornish *avel*. The mythical Isle of Avalon was the Isle of Apples. Whatever the tortuous routes by which this word has reached us today, in all its varieties, it hints at an origin in central Asia, just like apples themselves – and the horses that bore them on their original journeys.

Words for apples can be quite misleading. The most famous apple of them all, the mythical apple in the Garden of Eden, may not have been an apple at all. That sounds like an odd thing to say of something mythical, a storytelling device. But the original story didn't refer to an apple. The forbidden fruit growing in the Garden, which the serpent encouraged the woman to eat, was *tappuah*. This Hebrew word doesn't mean apple. In fact, it's only very recently that apple varieties have been developed which will actually grow in the hot, dry climate of Palestine, where the stories probably originate. Scholars debate what was actually meant by that word *tappuah*. It could have been an orange, pomelo, apricot or pomegranate – but it almost certainly wasn't an apple.

Homer's *Odyssey*, probably written in the ninth century BCE, apparently contains a reference to apples, growing in the orchard of King Alcinous, on the mythical Isle of Scheria:

> *Here luxuriant trees are always in their prime,*
> *pomegranates and pears, and apples glowing red,*
> *succulent figs and olives swelling sleek and dark.*

But these 'apples' – along with others in Greek myth, such as the one given by Paris to Aphrodite, or those growing in the garden of the Hesperides – could actually have been any round fruit. The Greek word *melon*, despite its shared roots with other Indo-European words for apple, is not specific – it could mean any plump, round fruit (including melons!).

It's not just the early words for apples that are tricky. Reaching Mesopotamia, the apple itself was discovered to be a bit of a trickster. By 4,000 years ago, when a type of apple close to our domesticated apples first appears in the Near East, people in this area had been farmers for thousands of years. They understood the ways of nature, and could control them ... but not when it came to fruit trees. It wasn't as though fruits and nuts hadn't been an important part of the diets of ancient people – these types of plants were just very hard to domesticate. Unlike the cereals and pulses, woody plants have more built-in genetic variability. Apples have two sets of chromosomes – like us. And they will *not* self-pollinate. They're described as 'highly heterozygous' – it's unusual to find the same version of a particular gene repeated on the other chromosome in the pair. In this way, they're also a bit like humans. Our heterozygosity (what a deliciously poetic mouthful of a word) means that children tend to differ from their parents. Similarly, fruit trees – and in particular, apples – don't stay 'true to type' – an annoying trick if you're a horticulturalist trying to breed trees which all retain a particular, desirable trait. The seedling progeny of a beautifully sweet-fruited apple tree will almost inevitably be too sour to eat – as the naturalist Henry David Thoreau wrote: 'sour enough to set a squirrel's teeth on edge'. But ancient apple orchardists eventually discovered how to make the apple stay true.

They found a way of capturing the qualities of a prized apple tree and passing on those traits to other trees. Four millennia ago, gardeners invented cloning.

Some plants lend themselves to cloning – they do it naturally. Any plant that spreads by sending out runners over or under the ground, to grow a new bit of itself, some distance away from the parent plant, is essentially cloning itself. You can divide the roots between the two of them and the youngster will carry on growing by itself, quite happily. My *Rosa rugosa* hedge is extremely good at cloning itself in this way. I have no doubt that, left to its own devices, it would propagate itself all over the place with its offshoots, and turn my whole garden into a thicket. I have to keep it in check by cutting back its runners and the brave new shoots attached to them. But if I wanted more *Rosa rugosa*, I could keep those shoots and shove them in the ground elsewhere – and they would take root and grow. Early agriculturalists were able to use this natural tendency of some plants to reproduce asexually to their advantage. They found that they could cultivate clones of figs, grapes, pomegranates and olives from cuttings, and that date palm would grow successfully from divided offshoots. But pears, plums and apples were much less malleable. They wouldn't stay true to type when grown from seed, and it was very hard to get cuttings to take root – especially in the dry lowlands of the Near East. There's plenty of evidence for wild and feral apples spreading by vegetative propagation, sending out suckers from roots, or sprouting anew from branches which have become covered with soil on the ground, but it seems more difficult to get domesticated apples to reproduce in this way.

Someone must have noticed, however, that trees can join together. This could have been an extremely ancient discovery. You can make a shelter from slender, living trees, bending them into a basic yurt-like frame. Even if you cut withies to make such a structure, they may take root and grow – especially if you're using willow or fig – and over time, the withies will merge and meld with each other where they cross. Having noticed that, perhaps having seen two wild trees growing too close and merging with one another, it's not too much of a leap of faith to ask: if I cut this tree and attach it to this other tree, could it grow? And it worked. Millennia before the transplant of a heart from one human being to another, our ancestors discovered that they could transplant fruiting limbs from one plant on to the rooted limb of another.

Grafting means that you can clone hundreds of apple trees from a single 'parent' (it's not a parent in the strict sense, but an identical sibling). It brings other advantages, too. If you plant a seed, you must wait years for it to grow and start flowering and fruiting. But if you graft a mature limb, or scion, even on to a juvenile rootstock, it will quickly start bearing fruit – you leapfrog right over the immature stage. You can add new cultivars to your rooted tree at any time. By choosing your rootstock carefully, you can affect the size of the tree too, making a dwarf out of a cultivar that would be a giant on its own. Some rootstocks bring advantages with them, which the variety you want to grow doesn't possess, such as pest or drought resistance. Grafting can also be used in another way – to save an ailing tree. If the roots have been attacked by pathogens, or the bark of the trunk has split, you can plant seedlings around your traumatised tree and encourage them to fuse with the trunk higher up, to carry essential water and nutrients up to the branches, from the soil, like a bypass graft.

Grafting seems like an astonishingly advanced technological development, and yet there's a hint it may have already been carried out on other plants by the time apples arrived in the Near East, in the early second millennium BCE. The clue comes from a snippet of cuneiform script on a fragment of a Sumerian clay tablet, dating to around 1800 BCE, discovered during excavations of the palace of Mari, in modern-day Syria. This ancient text referred to grapevine shoots being brought to the palace to be replanted. It's been widely interpreted as referring to grafting – but it's actually not clear whether those grapevine shoots were simply going to be used as cuttings: pushed straight into the ground. Vines do root easily, so perhaps that's a more likely scenario. Nevertheless, other clay tablets from Mari refer explicitly to apples being shipped in to the palace. The kings of Mari certainly knew the taste of apples, even if they weren't growing them or grafting them for themselves yet.

There's another text, or rather a collection of texts, dating to slightly later, which perhaps provides more dependable evidence for grafting. The Hebrew Bible consists of a compendium of stories and histories which were written and collected together over a thousand years, between 1400 BCE and 400 BCE, spanning the last centuries of the Bronze Age, into the Iron Age. Although grafting isn't referred to specifically, it's certainly implied, in several parables where cultivated vines reverting to wild forms are mentioned. It's highly likely that the Persians – whose empire stretched around the eastern Mediterranean and into India and western Asia – would have used grafting in their orchards, but there's no clear reference to this practice.

It's Ancient Greek literature that provides us with the earliest, unequivocal description of grafting. A passage from the Hippocratic Treatise, dating to the late fifth century BCE, runs: 'Some trees grow from grafts implanted into other trees: they live independently on these, and the fruit which they bear is different from that of the tree on which they are grafted.' The Romans planted orchards of sweet apples in Italy, along with cherries, peaches, apricots and oranges. By the time the Romans were becoming a force to be reckoned with throughout Europe, references to grafting abound. And it was the Greeks and then the Romans, through their trade networks, colonies and empires, who spread apples and orchards and the knowledge of grafting right across the continent. A wonderful mosaic from Saint-Romain-en-Gal in southern France, dating to the third century CE, shows the orchard year – from planting, grafting and pruning to harvesting and cider-making. For the Romans, cultivated apples were a sign of civilisation. Tacitus, writing in the third century CE, recorded the Germans eating *agrestia poma* – peasant, rustic or wild apples – in contrast to the *urbaniores* – the urbane, cultured, sophisticated fruit – that the Romans preferred. But, as the civilising influence of Rome spread across Europe, so did the cultivated apple. At least, that's probably what the Romans would have liked us to think. There's a chance, though, that cultivated apples arrived in Britain and Ireland much earlier than that. During excavations of Haughey's Fort, a late Bronze Age hillfort in County Armagh, Northern Ireland, archaeologists found what appeared to be a large, 3,000-year-old apple. It caused much excitement – but further inspection revealed it to be a puffball. A 3,000-year-old puffball.

So far, then, there's no direct evidence of pre-Roman apples in this corner of north-west Europe. But it's not such an extraordinary suggestion – there were extensive trade networks across Europe long before the classical civilisations took off. Indeed, some of the tin for making bronze may have come from Cornwall. Throughout Spain, France and Britain there are Celtic place names – which may have taken root in the Iron Age, if not before – that suggest the presence of apples in the ancient, pre-Roman landscape. From Avila in Spain, to Avallon, Availles and Aveluy in France, to the elusive Isle of Avalon or Ynys Avallach – somewhere in Britain – appley names could imply an even more ancient link to those ur-orchards in Kazakhstan. But this is still just conjecture. Those apple names could, presumably, just as easily relate to local, wild apples.

When Romans first arrived in Britain, Ireland, France and Spain, the native inhabitants would have been well used to exploiting their own local crabapples – just as they were in Germany. In a pit in Devon, archaeologists found a collection of baked-clay loom weights and an astonishingly well-preserved collection of pips, stalks and even whole wild crabapples, dating to the early Neolithic – nearly 6,000 years ago. Crabapples have also been found, threaded on to strings, at the 4,500-year-old tomb of Queen Puabi at Ur in Mesopotamia. But traces of apples have been discovered at even earlier archaeological sites, dating back to the Mesolithic, in Scotland. And in the Upper Palaeolithic site of Dolni Vestonice in the Czech Republic, crabapple remains have been found dating to around 25,000 years ago. It seems reasonable to assume that our ancestors have been eating wild apples for as long as there have been humans and apples in the same place. While crabapples seem to have formed part of the 'paleo diet', there were other uses for them – medicinally and in cider-making. And of course, crabapples are still used in their wild form. They're planted in orchards to help with pollination of cultivated apples. Their fruits are cooked and served with meat, or made into sauces and jellies – and are still used to make cider. And they're pretty. Alongside my own four *urbaniores*, I have a beautiful little *agrestia poma*, a pink-flowered, yellow-fruited, crabapple in my own garden.

The spread of sweet, plump cultivated apples, from the Near East, throughout Europe in the Bronze and Iron Age – carried out, if not exclusively, under the aegis of the Roman Empire – may be thought of as the first major diaspora. With the collapse of the Roman Empire, orchards were abandoned. But, in western Europe, apples survived in the gardens of monasteries, spreading across Europe again with the expansion of the Cistercian order in the twelfth century. In 1998, a single apple tree with red-golden apples was found growing on Bardsey Island – probably the last survivor of the monastic orchards there, and it's now being cultivated again. In eastern Europe, apples survived the fall of the Byzantine Empire in the eighth century, and were carefully curated and cultivated in the Muslim world. In the sixteenth, seventeenth and eighteenth centuries, the second major diaspora took place as European colonisers started to plant cultivated apples in the Americas, South Africa, in Australia, New Zealand and Tasmania. In 1835, when Darwin landed in Chile, he found the port of Valdivia surrounded by apple orchards. Tasmania would later become known as the 'Apple Isle' – an antipodean Avalon.

This second apple diaspora resulted in a great variety of apples – suited to diverse environments across the temperate zone. The success of apples in North America seems to have involved a 'return to the wild'. Apples were planted from seed and natural selection got down to work, weeding out individuals that could not thrive in the new habitat, with its harsh winters. New varieties emerged out of that sifting, whilst cultivars undoubtedly hybridised with native American crabapples too, borrowing useful local adaptations. The apple was able to remake itself, to fit its new habitat. It was out of this global spread of apples, and the winnowing of selection acting on seedlings once again, that our familiar modern cultivars began to appear in the nineteenth century. The McIntosh or 'Mac' (the namesake of the Apple Macintosh computer) emerged in 1811 in Canada; Cox's Orange Pippin appeared in 1830 in Buckinghamshire; Egremont Russet in 1872 in Sussex; Granny Smith in 1868 in Australia. In the twentieth century, selection became even more directed, precise and brutal. An incredible diversity was pruned back to just a few brand-names that would corner the global market in apples. And yet new varieties continued to be born

– of which some would become immensely successful: Golden Delicious appeared in West Virginia in 1914; Ambrosia in the 1980s in Canada; Braeburn in 1952, Gala in the 1970s and Jazz in 2007 in New Zealand.

Diversity has been reined in amongst our modern cultivated apples – but it's still impressive, especially compared with other species. The botanical expeditions of the late twentieth and early twenty-first century seemed to confirm what Vavilov had concluded on his 1929 visit to the orchards around Almaty – that all the huge variety we see in our modern apple cultivars could be traced back to those ancient ur-orchards of Kazakhstan.

Genetic revelations

Similarities in the shape of trees, and their flowers and fruit, together with clues from written history, all pointed to the foothills of the Tian Shan as the birthplace of domesticated apples, *Malus domestica*. In the 1990s studies of mitochondrial DNA – and chloroplast DNA, which is also inherited down the maternal line – confirmed the hypothesis that the Asian wild apple, *Malus sieversii*, was the ancestor of the modern, domesticated apple. There had always been the possibility that interbreeding with other wild-apple species had been instrumental in the development of domestic apples, but the geneticists uncovered what looked like an unbroken, fairly unsullied ancestral line going right back to the wild apples of Kazakhstan. It seemed that the DNA of our familiar eating apples was still, in the main, *Malus sieversii*. Given that this wild-apple species was so spectacularly variable in the wild, it didn't seem unfeasible that all – or nearly all – the variation seen in domestic apples could quite easily come from this single source. Some botanists even went as far as to put domesticated apples and the central Asian apple into the same species, *Malus pumila*.

But a huge new study of apple varieties published in 2012, led by French geneticist Amandine Cornille, has painted a new and unexpected picture of apple origins. This new analysis looked at large numbers of cultivars, from China to Spain, and used more complete samples of DNA than previous investigations – and revealed a surprisingly prodigious level of variation. Whereas most domesticated species contain just a fraction of the diversity seen in wild relatives, cultivated apples are just as diverse as most wild *Malus* species. But it was when Cornille and her colleagues drilled down into the variation, and carefully looked at comparisons between the domesticated apples and wild species, that they revealed the well-hidden secret of the apple. The geneticists found that cultivated apples did indeed come from the wild apples of Kazakhstan – but not exclusively. They discovered that cultivated apples had clearly bred with wild crabapples as the cultivars spread along the Silk Road. Apples had not emerged from a single geographic origin, over a short time period, but had continued evolving – and interbreeding with close cousins – over millennia. Throughout the apple's history – and despite the use of grafting to clone and genetically constrain apple populations – improvements have arrived on the scene through human selection of good-looking apples created by natural, open pollination. It's always been possible for wild cousins to add their own contributions to the gene pool. And those contributions just arrived – there was nothing intentional on the part of humans to orchestrate those crosses.

This interbreeding with wild apples wasn't just about adding details to the story of the domesticated apple; it subverted it. The original progenitor of cultivated apples was still *Malus sieversii* – with the origin of domesticated apples estimated to have occurred between 4,000 and 10,000 years ago. But the influence of other wild apples – particularly *Malus sylvestris*, the European crabapple – has been profound. The study showed that our modern, domesticated apples actually owe *more* of their genetic make-up to crabapples than to the original central Asian apples.

It's an extraordinary revelation, but echoes what has recently been found in quite a few other species – including other woody, domesticated plants such as grapes and olives. And it's much like the story of maize: cultivated *Zea mays* is genetically more similar to the highland wild varieties than to the originally-domesticated, lowland teosinte.

A few botanists had suggested in the past that cider apples might have been produced by crosses with crabapples, to introduce a desirable bitter, astringent quality. While the Cornille study showed that interbreeding had definitely occurred, it didn't show any difference between modern cider apples and eating apples. Both had roughly equal, considerable amounts of *Malus sylvestris* in their ancestry. If anything, the sweet, dessert apples had more. Geneticists have already begun to dig into exactly what these different sources of their heritage, of their DNA, means in practice. Genes for fruit quality have been inherited – and preserved – from that original progenitor, *Malus sieversii*. In contrast, genes from local wild species have contributed adaptations to the new environments that the domestic apple found itself growing in, as it spread from the forests of the Tian Shan.

The 2012 Cornille study also showed plenty of gene flow in the opposite direction, from cultivated apples into wild ones. So domestication has influenced the evolution of apple species that haven't even been domesticated (just as it's done with horses and wolves). This evidence – of gene flow in both directions – has emerged so recently that agronomists and conservation biologists are still trying to get their collective heads round the implications. What about the wild apples – are they threatened by the penetration of domestic genes into their genomes? This exchange of DNA isn't a new phenomenon – it must have been happening since the dawn of domestication. It's easy to leap to conclusions, and imagine that all introgression from domestic into wild species is deleterious and undesirable. But it's also possible that some domestic genes could be beneficial. We need to invest in answering these questions in order to direct our conservation efforts and protect wild species to the best effect. Conservation seems morally right, and an altruistic endeavour, but we also have more selfish reasons to care about the health of wild species. Genetic analyses of our modern apple cultivars show that some of them seem to be dangerously closely related: the equivalent of second or even first cousins – or sometimes even siblings. This increases the chance of genetic diseases becoming serious – because it brings rare variations in genes together. The genetic diversity across modern, domesticated apples looks impressive when compared with other tamed species – there is no detectable domestication 'bottleneck' – but the total diversity hides a more troubling reality. Apple production is based on cloning. There might be plenty of genetic difference *between* clones, but absolutely none within them. Although there are millions of domesticated apple trees growing in the world, these actually represent just a few hundred clones – a few hundred actual individuals. Some are fruiting scions, some are rootstock. It means that apples are quite seriously threatened by alterations to their environment – such as the emergence of new pathogens and changing climate.

So it's even more important to maintain a healthy pool of genetic diversity amongst wild apples – we may well need to tap into it to keep our domesticated apples healthy. In fact, we undoubtedly will. Wild apples might also hold the key to tackling some of the common problems that are already affecting our domestic apples. Botanists visiting the wild-apple forests of Kazakhstan have noticed that some of the trees seemed to be unscathed by canker and scab – apparently resistant to these diseases. Some also appeared to be able to grow in extremely arid conditions – a drought-resistant trait which could be immensely useful for some cultivated apples. It's also clear that whatever happens in the lab, this needs to be supported by expeditions into the field. We still need the Vavilovs, the Forslines, the Junipers to trek out into the ancient places, the wild places, and bring back their precious samples. Out there in the wilderness may exist the genetic answers to some of the challenges facing our orchardists today, and some problems that we haven't even thought of yet.

Genetics is shining a bright light on ancient origins of many species. We get clues from archaeology and history, but sometimes these clues can be misleading. The evidence is always patchy. Interrogating DNA, both modern and ancient, gives us the chance of filling in some of the gaps, by offering us another perspective on the past. As sequencing entire genomes becomes easier and quicker, we're now gathering in armfuls of unexpected insights into the histories of our domesticated species, from the astonishingly ancient origin of domesticated dogs – to amazingly early traces of wheat in Britain; from identifying lowland Balsas teosinte as the original ancestor of maize – to the real crabapple nature of apples. But one of the most surprising genetic revelations of all has involved a very familiar species: *Homo sapiens*.

10

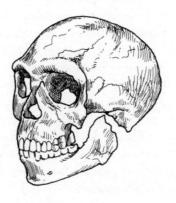

HUMANS

Many historical problems can be understood only because of the interaction between man, animals and plants.

Nikolai Vavilov

Professor Busk's Grand, Priscan, Pithecoid, Mesocephalous, Prognathous, Agrioblemmatous, Platycnemic, wild Homo calpicus

In 1848, a skull was discovered during British mining operations at Forbes Quarry on the north face of the Rock of Gibraltar. It was presented to a meeting of the local Gibraltar Scientific Society, but no one could make sense of this odd, heavy-browed, thick-set cranium with its gaping eye sockets. It was left on a shelf to gather dust.

Eight years later, another skull was found, along with some bones, in another quarry – this time in Germany. The remains were unearthed in the Feldhofer Grotte of the Neander Valley near Dusseldorf. Workmen clearing the mud from caves prior to quarrying found what they thought were cave-bear bones – but a local teacher recognised that they were human and collected them up. Professor Mayer, at Bonn University, suggested that the bones may have belonged to a Mongolian military deserter, who had died from rickets, frowning in agony with the pain and forming heavy brows in the process. But Professor Schaaffhausen, at the same university, thought that the Feldhofer skull and bones were normal – not pathological. As the remains had been found alongside the bones of extinct animals, he reasoned that this human must have been a very ancient inhabitant of Europe. In 1861, London anatomist George Busk translated Schaaffhausen's paper on the Feldhofer fossils – he agreed that the skull was potentially that of an ancient type of human, and he called for more fossils. The next year, the Forbes Quarry skull was duly packed up and sent to London.

Come 1864, Busk published his report on the 'Pithecoid Priscan Man from Gibraltar', in which he claimed that it resembled 'the far-famed' Feldhofer fossil. The remains from Gibraltar and the Neander Valley weren't mere oddities, he argued, but representatives of a lost tribe that had roamed 'from the Rhine to the Pillars of Hercules'. Darwin saw the 'wonderful Gibraltar skull' in that same year, but he neglected to make any further comment on it. Busk's friend, Hugh Falconer, wrote to him on 27 June, suggesting a name for the specimen:

My dear Busk,

A hint or two about the names which I have been rubbing up for the Priscan Pithecoid skull, Homo var. calpicus, from Calpe, the ancient name for the Rock of Gibraltar. What say you?

... Walk up! ladies and gentlemen. Walk up! and see Professor Busk's Grand, Priscan, Pithecoid, Mesocephalous, Prognathous, Agrioblemmatous, Platycnemic, wild Homo calpicus of Gibraltar ...

Yours ever, H. Falconer

But Busk wasn't quick enough off the mark. Just a few months after Busk's 'Pithecoid Priscan Man' was published, geologist William King, of Queen's College, Galway, got his hands on a cast of the Feldhofer skull. He recognised it as an ancient type of human, but – rather than an archaic sort of *Homo sapiens*, he thought its peculiarity warranted giving it a new species name. He proposed *Homo neanderthalensis*, after the German valley. And so it was King, not Busk or Falconer, who became the very first person to name a species of ancient human – and, of course, the name stuck.

Busk moved on to studying extinct hyenas and cave bears. Falconer died in 1865. And the Forbes skull was tucked away on a shelf again, this time in the Royal College of Surgeons. If events had played out a little differently that year in 1864 – if Busk had been a little less cautious, perhaps – we'd be talking about Calpicans today, not Neanderthals.

Since that first discovery and the recognition that other human species had existed, fossils kept turning up, often in unexpected places, and more and more names were added to branches populated by ancient species more closely related to us than to any other apes – together, we are the hominins. There are now more than twenty named species of hominin, including eight species that have existed within the last 2 million years and that are closely related enough to us to include in our own genus, *Homo* – the humans.

Neanderthals – the first to be named – remain central to discussions about human origins. Thousands of bones have now been found from more than seventy different sites. And what look like characteristic Neanderthal stone tools have been found at hundreds of sites. For a long time, they've been thought of as close cousins of ours. They behaved in similar ways to modern humans living at the same time – they knapped stones to make their hunting weapons and to form scrapers and knives, they buried their dead, they collected shells, used pigment and made marks on the walls of caves. These other people, this 'lost tribe', coexisted on the planet with modern humans for thousands of years. But then they disappeared. The enduring question has been – did we meet? Were Neanderthals another ancestor, or just close cousins and – ultimately – a dead end on our family tree?

For years, palaeoanthropologists and archaeologists have argued over the fate of Neanderthals – and in particular, whether modern humans and Neanderthals had ever mixed. Some skeletons appeared to show signs of interbreeding, with classic Neanderthal traits appearing in what otherwise looked like a modern human skeleton. But many experts remained unconvinced. The resolution to the question has had to wait until our modern technology developed to a point where it could provide access to the answer. With new technology, we have the ability to extract and sequence DNA from ancient bones. And finally, it seems that it may be possible to answer the question: did our *Homo sapiens* ancestors interbreed with *Homo neanderthalensis*? Are *we* hybrids?

The unfolding story of human origins

The history of research into human origins follows a course with which we're now well acquainted. The story began to be pieced together from the study of living humans across the world, and during the nineteenth century there was much discussion over whether humanity could be divided into different races, or even separate species, and if so, whether these had separate origins. As ancient, fossilised bones from earlier types of humans and proto-humans were discovered, starting with those skulls from Gibraltar and Germany, but soon including even earlier fossils from Africa, these had to be fitted into the story too. During the twentieth century, the great debate rumbled on: had modern humans, *Homo sapiens*, arisen from multiple origins – in Africa, Europe and Asia – or a single one?

The multiregional model proposed that we'd evolved from earlier species across a huge geographic range, with widely separated populations across several continents somehow remaining united through gene flow between them. In contrast, the Recent African Origin or 'Out of Africa' model proposed precisely what its name suggests: the emergence of *Homo sapiens* in a much more discrete geographic region, followed by an expansion of our species across the rest of the Old World, eventually reaching the New World.

In 1971, a determined 22-year-old student called Chris Stringer set off in his ancient Morris Minor to drive across Europe, tracking down fossil skulls in museums. He was armed with measuring instruments – protractors and calipers. He'd written to institutions that he knew possessed ancient skulls, and he'd had some letters back. But with others that he'd heard no word from, he just had to take a chance, and hope that he'd be granted access when he arrived. In the end, Stringer covered 5,000 miles and managed to collect measurements from fossil skulls unearthed in Belgium, Germany, what was still Czechoslovakia back then, Austria, Yugoslavia, Greece, Italy, France and Morocco. He returned to Bristol with all his data, which he analysed using a powerful statistical technique – multivariate analysis – which allows many measurements to be compared at once. He was keen to compare Neanderthal skulls with those of early, modern human Europeans, or Cro-Magnons, from around 30,000 years ago. He hoped to be able to answer that burning question: did Neanderthals evolve into Cro-Magnons, or were they a separate species?

What Chris Stringer found, by comparing the measurements of all those fossil skulls, was that Neanderthals certainly looked like a separate, distinct branch on the human family tree – and they appeared to have evolved in Europe. Cro-Magnons, on the other hand, were clearly part of *Homo sapiens*, modern humans, and seemed to have suddenly arrived in Europe, rather than to have evolved there *in situ*. Some scientists had by now suggested that modern humans could have interbred with Neanderthals, in the Middle East or Europe, but Chris found no evidence from the fossils of any interbreeding between the two species.

Chris Stringer wrote up his PhD, having made a significant contribution to some of the big questions around human origins, but what he couldn't tell from the fossils he'd studied was *where* modern humans had come from – where they'd first evolved. In 1974, Chris got a chance to look at a skull from Omo-Kibish in Ethiopia, which had been discovered by a team led by Richard Leakey, in 1967. At the time, the skull was estimated to date back to around 130,000 years ago. Many scientists back then thought that *Homo sapiens*, as a species, was only around 60,000 years old. But when Chris looked at the Omo-Kibish skull, it didn't appear to be from an archaic species. With its small brow ridge and domed, rounded braincase, the skull looked modern – a good candidate for an ancestor of the

Cro-Magnons of Europe. And with a date this early, it suggested that our species may have arisen in Africa.

Over the next decade, more evidence accumulated in favour of a single African origin of our species. By 1987, genetics had joined the debate, with a groundbreaking paper published in the prestigious journal *Nature*. Three geneticists from the University of California – Mark Stoneking, Rebecca Cann and Allan Wilson – examined the mitochondrial DNA of 147 people from all around the world, and they constructed a phylogenetic family tree from the data. The tree was firmly rooted in Africa. Over the next couple of decades, the genetic data began to pile up, taking in whole genomes, from many more living people, and it all seemed to point to an African origin. That continent contained the most genetic diversity – some 85 per cent of the global total, in fact – a good sign that this was our original homeland. The Omo-Kibish skull itself was re-dated, pushing it even earlier, to almost 200,000 years old. Based on the differences in genomes of modern humans, including living people and ancient ancestors, geneticists have suggested an even earlier divergence date of 260,000 years ago. And in the summer of 2017, there was another revelation as human fossils from Jebel Irhoud in Morocco were dated to between 280,000 and 350,000 years ago. There are several skulls from this site and while the shape of the braincase amongst these is archaic – long and low – the faces are small and pulled back under the braincase: a definitive characteristic of modern humans.

So the picture that had emerged over the forty years following Richard Leakey's discovery of the Omo-Kibish skull, and over the thirty years since the first mitochondrial DNA studies into human origins, was one of a broad origin across Africa and perhaps slightly beyond, though not as extensive as the Old-World-wide base of the original multiregional theory. But then, some time after 100 thousand years ago, modern humans started to spread beyond that homeland around the globe. They spilled out of Africa, firstly into Arabia and from there round the coast of the Indian Ocean until, by around 60,000 years ago, there were modern humans in Australia. Between 50,000 and 40,000 years ago, modern humans spread westwards into Europe.

But our ancestors weren't the first humans to live in Europe or Asia. *Homo erectus, Homo antecessor, Homo heidelbergensis* and *Homo neanderthalensis* had lived there for hundreds of millennia before our ancestors arrived on the scene. Most of these other species had gone extinct before the moderns arrived. But not Neanderthals. Their population may have been dwindling, hit hard by two particularly nasty downturns in climate in the run-up to the peak of the last Ice Age, but they hung on – until their eventual disappearance from the fossil record some 30,000 to 40,000 years ago.

Through the 1990s and into the 2000s, the debate continued about whether there had been any interbreeding of modern humans with Neanderthals. A few fossils were put forward by some palaeo-anthropologists as providing evidence of hybridisation, but still, most experts in the field were unconvinced. Although careful dating of fossils suggested that modern humans and Neanderthals had indeed existed at the same time and in the same general area, in the Middle East and in Europe – with potentially thousands of years of overlap – it seemed that the two populations had always stayed separate. Curiously separate, even. Mitochondrial DNA extracted from Neanderthal fossils was distinct from that of modern humans – with a divergence date estimated at around half a million years ago. Early studies of nuclear DNA from Neanderthal chromosomes suggested a similar date for a last common ancestor of modern humans and those archaic Europeans. And there didn't seem to have been any mixing of the populations after that split.

Then, in 2010, a group of geneticists working at the Max Planck Institute for Evolutionary Anthropology in Leipzig published an astonishing finding. They'd extracted and analysed DNA from fragments of Neanderthal bone dating to more than 40,000 years ago, from a cave in Croatia. This time, they'd looked more comprehensively at the nuclear genome. And they could compare the draft Neanderthal genome they managed to assemble with genomes of living, modern humans. This comparison revealed that some living people – those of broadly Eurasian ancestry – had more in common with Neanderthals than those with largely African ancestry. The most likely explanation for this inconsistency was that the ancestors of some of us had interbred with Neanderthals. It was an incendiary suggestion. Plenty of scientists published counter-arguments. But as more ancient DNA was recovered from fossils, and compared with the DNA of living humans, it became harder and harder to reject the evidence for interbreeding. Just as *japonica* rice interbred with proto-*indica* as it travelled west, and the wonderfully plump apples from Kazakhstan had interbred with European wild apples as they spread through Europe, our modern human ancestors had interbred with the indigenous humans of Europe and western Asia: the Neanderthals.

The development of new genetic tools – ways of analysing DNA inside mitochondria (and inside chloroplasts in plants) and then within the chromosomes themselves – has both enabled and constrained our understanding of the past. Or at least, the early studies were constraining. Both mitochondrial and chloroplast DNA provided a simple and single-rooted path back through history: each is inherited only through the maternal line – although informative, in some ways, each is equivalent to a single genetic marker. There was always a very good chance that this view of history would be unrepresentative of the whole picture. It's based on such a small fraction of the DNA that cells contain. We need to trace back the evolutionary histories of each gene – and not only that, the stretches of DNA between them that impact on how those genes are read and expressed – before we can truly access the wealth of historical knowledge contained in this amazing biomolecular library. The history of genetic analysis itself has forced a particular trajectory on to the way our understanding of the origin of many species has unfolded – our own included.

So here's what we know now (though some of this will change as more data comes in): our species originated – probably across a vast, connected area – in Africa (and perhaps even extending into western Asia). Although there may have been earlier migrations, one major migration, starting between 100,000 and 50,000 years ago, led to the colonisation of the rest of the globe. And our ancestors definitely interbred with other, archaic human species or populations. So although the idea of a recent African origin for all of modern humankind has stood up, it's certainly become blurred around the edges.

Those are the headlines. But the detail is even more fascinating.

Rather than having a well-defined centre of origin, our species may have originated in a diffuse, continent-wide manner – something sounding perhaps more multiregional in nature, but not on a global scale. Much like wheat – where the conceptual origin of einkorn and emmer focused in on the Karacadag Mountains of south-eastern Turkey for a while, then embraced the whole of the Middle East – the imagined homeland of modern humans narrowed, to a discrete area of Africa, then broadened – to encompass all of Africa, and perhaps even a bit of Asia too. Various evidence – both genomic and palaeontological – has been put forward to argue for an origin of modern humans in eastern, central or southern Africa. But perhaps we don't need to choose between them. Modern characteristics may have arisen in a mosaic, piecemeal fashion, spread across populations – connected by gene flow – right across Africa, and a little beyond. African DNA contains echoes of a complex history, with traces of ancient migrations across sub-Saharan Africa, deep splits between populations – but also of mixing together. For tens of thousands of years, *Homo sapiens* was largely restricted to the continent of Africa – but then the population began to expand, and spread.

The latest, most comprehensive genome-wide analyses support a single, major migration out of Africa, between 50,000 and 100,000 years ago, leading to the colonisation of the rest of the globe. After leaving Africa, the pioneers split into one wave headed eastwards, along the coast of the Indian Ocean – eventually reaching south-east Asia and Australia. Another wave headed north and west, into western Asia and Europe. The eastbound colonisers may have met with modern humans whose ancestors had emerged from Africa during an even earlier migration, reaching all the way to Australia and Papua – as it stands, the fossil record of south and south-east Asia is so thin on the ground that it's impossible to rule out the existence of a very early eastwards migration.

The date of hybridisation with Neanderthals, the long-time European indigenes, is estimated to be between 50,000 and 65,000 years ago – soon after the out-of-Africa dispersal of modern humans began. Non-African people contain a small percentage of Neanderthal

DNA – on average, 2 per cent – whereas people of generally African heritage have little to no Neanderthal DNA knocking around in their genomes. Having had my DNA tested, I am myself, apparently, 2.7 per cent Neanderthal. So I'm not 'pure' *Homo sapiens*. (Nobody is. In fact the whole idea of 'purity' in species and subspecies turns out to be an illusion, a hangover from the nineteenth century perhaps, that modern genetics has finally put to rest.) East Asian people tend to have a little more Neanderthal DNA than west Asians and Europeans. There are several possible reasons for this. The ancestors of east Asians – after splitting away from the western Eurasian lot – may have interbred more enthusiastically with Neanderthals. We also know that Neanderthal DNA has, largely, been weakly selected against since it first arrived in modern human genomes. So it's possible that the ancestors of both western and eastern populations started out with the same amount of mixed-in, introgressed Neanderthal DNA, and then natural selection could have weeded out more from western Eurasian genomes. Lastly, the lower level in the west may be due to a dilution effect – through interbreeding with populations that didn't have any Neanderthal DNA, possibly from North Africa.

But it's not just Neanderthals that our modern human forebears were hooking up with. In the genomes of modern-day people from East Asia, Australia and the Melanesian islands in the south-west Pacific, there are traces of interbreeding with yet another archaic population. In Melanesian genomes, 3–6 per cent of DNA comes from another type of ancestor – known only from a finger bone and a couple of teeth from Denisova Cave in Siberia. We don't know what these people looked like – the fossil evidence is so meagre. But what we do know, from looking at the ancient DNA extracted from that bone and those teeth, is that they were neither modern humans – nor Neanderthals. There's just not enough in the way of fossil evidence to give these people their own species name, so for now they're simply known as 'Denisovans'. The interbreeding between modern humans and Denisovans probably happened in Asia, before Australia and the Pacific Islands were colonised. There's also evidence for interbreeding with other, archaic – though currently unidentified – species within Africa. Modern African genomes harbour these memories of other ancient humans, even if we don't yet have any fossil evidence to link up with those genetic ghosts.

Genomics – the study of entire genomes, not just bits of DNA parcelled up in mitochondria, or individual genes laid out in our chromosomes – has revealed a rich and complex history that we had no idea about, just ten years ago. Our ancestors met up with a range of other people – people different enough to be considered as separate species – and joined them, interbred with them. As American palaeoanthropologist John Hawks wrote in his blog: 'It is notable that we now have evidence for interbreeding among every kind of hominin we have DNA from, and some we don't.' Geneticist and author Adam Rutherford, always focused on an exquisite turn of phrase, has described the dalliances that led to the inception of humanity as we know it as 'one big, million-year clusterfuck'. Humans have always been – as Rutherford so neatly encapsulates it – both 'horny and mobile'.

As well as providing clues to the origin of *Homo sapiens*, to the original colonisation of Eurasia, and those astonishing revelations about interbreeding with other human species, genomes also contain traces of later events in prehistory. Hidden deep in our DNA are memories of countless voyages and expeditions – of pioneers and explorers whose names are long forgotten. It's such a densely overwritten palimpsest, but geneticists are finally managing to draw out some of the detail from that archive.

In Europe, there are genetic echoes of three major waves of immigration. The first wave represents the Palaeolithic colonisers – although the very first of this group to arrive, reaching Britain at the far western edge of Europe by 40,000 years ago, leave little genetic trace. Their population would have crashed at the last glacial maximum. But after the ice sheets receded, survivors in southern, Mediterranean refugia recolonised the north. These hunter-gatherers, though still nomadic, became a little more settled as the climate improved – as we've learnt from Mesolithic sites like that of Star Carr in Yorkshire. These people were soon to be joined by the second major wave of incomers – who would bring a whole new way of life with them. Farmers, originally from Central Anatolia, expanded across Europe, in fits and starts, and probably travelling by boat, reaching the Iberian Peninsula around 7,000 years ago, and settling in Scandinavia and Britain by 6,000 years ago – 2,000 years after that genetic trace of wheat at the bottom of the Solent. Rather than completely replacing local, hunter-gatherer populations, genetic studies reveal that the farmers joined forces with them. The Neolithic had arrived. In some places, foragers quickly switched from hunting and gathering to settling down and farming. In others, such as Iberia, people continued to hunt alongside farming. The third wave of immigrants arrived, with their horses and their new language, in the early Bronze Age – around 5,000 years ago – as the Yamnaya population expanded and spilled over into Europe. If you have broadly European heritage, you may well still have the odd piece of DNA from these ancient horse-riders and herders in your genome – a little bit of Yamnaya tucked away – despite all those generations and dilutions of that DNA in between. Sadly, this doesn't mean that you have a natural affinity with horses, or an innate ability to ride them – that's still something that needs to be learnt!

The horse-riding herders of the steppe pushed east as well, replacing hunter-gatherer populations in southern Siberia. And another west-to-east migration in Asia took place around 3,000 years ago. Winding back much earlier, genetic studies have also helped to answer questions about the colonisation of the Americas. During times of low sea level, north-east Asia was connected to North America by the land bridge of Beringia. Human colonisers crossed it, to gain a toehold in the Yukon, before the peak of the last Ice Age. But they were stuck there until the huge ice sheets that blanketed North America began to melt a little at the edges, around 17,000 years ago. Then they could start to roam further south, probably using boats to travel and settle along the Pacific coast – getting as far as Chile by 14,600 years ago, as shown by the site of Monte Verde. All this we know from archaeology, but there were also some important challenges to this story. The skulls of some early Americans seemed to show morphological links with Polynesian, Japanese and even European populations. It was suggested that there had been an early migration into the Americas, and this population had later been replaced by colonisers from north-east Asia and Beringia. But when ancient DNA was extracted from those bones, it proved to be closest to living Native Americans, and the next closest match was Siberian and East Asian DNA. The idea of a population replacement could finally be put to rest – the first colonisers had come over the Bering land bridge, from north-east Asia, and filled up the continents, from north to south. However, there are genetic traces of significant later migrations in the far north – eastwards expansions of circumpolar people, from north-east Asia to the icy northern reaches of North America and into Greenland. The first was a Palaeo-Eskimo migration around 4,000 to 5,000 years ago, followed by an Inuit expansion between 4,000 and 3,000 years ago.

In Africa, genomes of living people also bear witness to great population shifts – ancient expansions and migrations. Some 7,000 years ago, Sudanese pastoralists migrated to central and eastern Africa; 5,000 years ago, Ethiopian agro-pastoralists expanded into Kenya and Tanzania; and a major expansion got under way 4,000 years ago, when Bantu-speaking farmers spread from their homeland in Nigeria and Cameroon, southwards. They replaced foragers as they went, pushing them into marginal habitats, where we find the last remaining hunter-gatherers, such as the Bushmen of Namibia, eking out an existence today.

Sunshine, mountain-tops and germs

As humans spread around the world, and as the climate fluctuated, they faced new challenges. Our ancestors adapted in various ways. Some adaptations would have been physiological – adjustments occurring within a lifetime; whereas others involved genetic alterations – the real stuff of evolution. Together, these changes allowed people to survive and thrive in testing environments. As humans made their way into northerly latitudes, they would have found themselves entering landscapes where the environment transformed itself around them – as the seasons turned. Summers brought long days, while winter days were short, with sunshine a rare commodity. And as far as your body is concerned, it really *is* a commodity. Sunny days not only lift the spirit, they provide us with a metabolic benefit – because when you're out in the sun, your skin is busy making vitamin D. Or at least, we transform a cholesterol-based compound to something that's *almost* Vitamin D in the skin, then the liver and the kidneys carry out the last steps, adding hydrogen and oxygen to activate the vitamin.

The importance of vitamin D to the body was elucidated in the early twentieth century, as researchers sought to understand and cure a condition which caused skeletal deformities in children: rickets. The industrialisation of Europe may have been a great technological step forward, and it may have ultimately led to all sorts of improvements in people's lives, but there were plenty of casualties along the way. Crowded cities, working in factories, skies full of smog – all left their mark on the children of the Industrial Revolution. They didn't grow properly, and their soft young bones set in awkward curves. Rickets continued to be both a troubling affliction and a mystery, until 1918, when a British physician by the name of Mellanby found that he could give dogs rickets by keeping them indoors and feeding them porridge – and that he could reverse the effect by giving them cod-liver oil. The following year, a German researcher called Huldschinsky found that shining ultraviolet light on to children with rickets could cure them. Other studies found that all sorts of food treated with ultraviolet light – vegetable oil, eggs, milk and lettuce – could protect against this disease. Without knowing it, the researchers were converting the cholesterol and plant sterols in these foods to pre-vitamin D. When, at last, the chemical identity of the essential compound was uncovered, chemists could begin to artificially synthesise vitamin D: finally, there was a pill for rickets. The German chemist who made the breakthrough, Windhaus, won a Nobel Prize for his efforts in 1928.

Yet it still wasn't clear how this substance worked its magic on bones. Over the ensuing decades of the twentieth century, research focused on following the compound on its journey through the body. It revealed that the vitamin acted like a hormone – once activated by the kidneys, it travelled in the bloodstream to the gut, carrying a message: 'Get calcium'. But vitamin D is a busy little chemical, and by the 1980s it was becoming clear that, as well as its important job in calcium metabolism and building bones, it also plays a crucial role in the immune system. A lack of vitamin D means that you're more likely to develop autoimmune diseases (where the armies of your immune system start to engage in friendly fire, or even mutiny), including diabetes, heart disease and specific types of cancer. A tiny dose of about 30 nano-grams per millilitre of blood seems to be the minimum amount of vitamin D the body needs to function healthily. While you can get some in your diet, most of us get around 90 per cent of the vitamin D we need by making it, in our skin, in the presence of sunlight.

Of course sunlight, particularly the ultraviolet rays within it, is also potentially damaging. Human skin contains several compounds which act as natural sunscreen, including the pigment melanin. If you're exposed to more sunlight than usual, your skin will start to make more of it, and you tan. This isn't only true for pale-skinned people – dark-skinned people tan too. The first modern humans entering Eurasia would probably have had dark skin – perfect for the climate they'd come from. In very sunny places, you need lots of melanin to avoid sunburn – so it's easy to see why natural selection would favour darker skins in equatorial regions. At the same time, in the tropics, enough UV radiation will make it through that filter to allow skin to photosynthesise vitamin D. But in a place with less sunlight, it seems logical to assume that dark skin could be so effectively filtering out UV that it would become impossible to make enough of this important vitamin. The adverse effects of deficiency, from an impaired immune system to rickets, would then mean that a selective pressure would be operating: anyone with slightly paler skin is likely to have had an advantage in survival and reproduction – they'd be more likely to pass their genes on to the next generation. And so whenever a chance mutation arose that tinkered with melanin production and produced paler skin, it would tend to spread through the population. It seems that this is exactly what happened. As you go further north, skin colour gets progressively paler. Both northern Europeans and northern Asians went through this process of adapting to low levels of sunlight separately – but via different mutations. It looks like a classic case of convergent evolution – a similar outcome achieved by different means.

The 'vitamin D hypothesis', positing that pale skin evolved as an adaptation to the lack of sunlight in northerly latitudes, seems to make a lot of sense. The observation that dark-skinned people today, in the UK and in North America, tend to be subject to vitamin D deficiency more often than their pale-skinned counterparts, appears to support the hypothesis. However, careful measurements of vitamin D levels in living people have thrown a spanner in the works. Studies tracking vitamin D levels and exposure to sunlight have produced interesting and unexpected results. The studies found that vitamin D levels increased (up to a point) as exposure to sunlight increased, just as predicted. Covering up with clothes was understandably associated with lower levels of vitamin D in the bloodstream. But thinly applied sunscreen – while protecting from sunburn – didn't appear to reduce vitamin D production. And neither did having a darker skin colour. Rather surprisingly, there was no difference in the boost to vitamin D production measured in dark-skinned versus fair-skinned people, exposed to the same amount of sunlight.

This research certainly suggests that people with dark skin seem to be able to make vitamin D just as efficiently as pale people. At first glance, these new findings look as though they might bring all our theories about the evolution of human skin colour tumbling down. But there are still some real observations that require explanation: the skin colour of indigenous people *does* get paler as you go further north, and people with dark skin *do* tend to suffer more from vitamin D deficiency in northern countries.

The first observation leads to a question about how changes happen in evolution – and they don't always happen because a certain muta-tion confers an advantage. Sometimes they occur as mutations that are nearly neutral in terms of selection pressure spread through a population, in what's known as genetic drift. This is essentially a random process, owing a lot to chance. Perhaps what happened as our ancestors moved north was that a strong selection pressure for dark skin – as defence against sunburn and skin cancer – eased off. Then, mutations for paler skin could occur without being weeded out – and could end up spreading via genetic drift. And actually, there isn't a steady gradient of paling skin from the equator to northern latitudes; fair pigmentation evolved – probably quite late – only in populations living in the far north of Europe and Asia. The rest of Europe and

Asia is full of people whose skin colour isn't linked to latitude. Another problem with the vitamin D hypothesis is that there aren't many skeletons bearing evidence of rickets prior to the Industrial Revolution.

But what about people with dark skin today, in the UK, in North America, and the problem of vitamin D deficiency? One of the studies in contemporary populations found a clue, by asking people to fill in detailed questionnaires about what they *did* when it was sunny. It turns out that pale-skinned people tended to rush out in the sun when it appeared, while dark-skinned people more often stayed inside. In a place with plenty of strong sun, that's probably a good strategy – but in the less sunny north, with weaker sun and less of it, you really need to make the most of sunny days – particularly in winter. For early modern humans – Palaeolithic, nomadic hunter-gatherers – spending time outdoors (or, more realistically, outside your tent) on a daily basis, all year round, would have been an inevitability. So while dark skin may represent an adaptation to strong sun in equatorial regions, the converse – pale skin as a general adaptation to more northerly latitudes – doesn't stand up to scrutiny. Under the skin, however, there are less obvious changes in vitamin D metabolism that may yet represent real adaptations to higher latitudes. Northern European genomes contain mutations that increase the levels of the vitamin D precursor in the body, while dark-skinned people possess other mutations that enhance the absorption and transport of vitamin D around the body. Yet again, a simple, pervasive hypothesis has given way to a much more complex picture, as epidemiological rigour and genomic data are brought to bear. Over recent years, the story of human adaptedness to different latitudes has become more interesting and much less straightforward; much less – you might say – black and white.

Whilst changing latitude appears to be linked to some metabolic adaptations, high altitudes also present a particular challenge. A specific variant of a gene called EPAS1 has been connected with an ability in some humans to handle low oxygen levels at high altitude. It's linked to a reduced level of haemoglobin production, perfect for an oxygen-poor environment, along with denser networks of blood vessels. The EPAS1 variant shows clear signs of having been selected for in Tibetans – but its origin was mysterious. The pattern didn't fit with it being either an existing variant that suddenly came into its own as people started to live at higher altitudes, or with it being a new, fortuitous mutation. Where had it come from? It was absent in all the individuals who provided DNA samples for the ambitious, international '1000 Genomes' project, which was completed in 2015, apart from just two Chinese people. But – it was there in the Denisovan genome. So it looks like this EPAS1 variant in modern Tibetan genomes has been inherited from Denisovan ancestors – and then jealously preserved by positive selection. Like apples acquiring useful new adaptations by interbreeding with crabapples, our ancestors picked up on local, genetic knowledge.

One of the most significant challenges of new or changing environments is the presence of novel pathogens. We're constantly fighting a battle against microbes, and the history of this evolutionary arms race is embedded in our genomes. Some of the genetic variants that have entered modern human genomes quite clearly came from Neanderthal and Denisovan ancestors – presumably they conferred some protection against specific infections, at particular times and in particular places.

A gene inherited from Neanderthals that's involved in fighting off viral infections pops up in one in twenty Europeans, but appears in over half of the modern population of Papua, where it seems to have been strongly selected for. Other genes linked to the immune system also seem to have come in from Neanderthals, and to have been selected for more strongly in some populations than others. It's in patterns like this where we see the crucial importance of contingency in evolution: a genetic variant that may confer some resistance to a pathogen will become important – and selected for – if populations are exposed to that pathogen. If not, the variant may well disappear, or at least drop to low frequency in the gene pool.

There's a whole bunch of closely related genes in our genomes that are all involved with the important task of helping the body to recognise foreign invaders, and to mount attacks against them. They're also involved in self-recognition – they encode proteins which are stuck like flags on the surface of our own cells, so that the immune system doesn't mistake them for alien pathogens. They're called HLA genes, and it's estimated that, in modern Eurasians, more than half of these genes have been inherited from Neanderthals or Denisovans.

There is a downside to some of the genes we've inherited from archaic humans, however. While they may have proved useful at various times in the past, some alleles are linked to deleterious effects today. Certain variants of HLA genes may create a predisposition to developing autoimmune diseases. This is essentially a failure of the self-recognition role of these HLA genes: the flag looks odd – alarmingly foreign to the immune system – and it ends up launching an attack on its own body's cells. The immune system gene HLA-B*51, inherited from Neanderthals, is associated with a higher risk of developing an inflammatory condition called Behcet's disease, which causes ulceration of the mouth and genitals, and inflammation of the eyes that can eventually lead to blindness. It's rare in the UK, but affects about one in 250 people in Turkey. Behcet's disease is also known as the 'Silk Road disease', but its origins appear to be much more ancient than the human trade in cloth. Millennia before the pathways we know as the Silk Road operated as trade routes, they were important for migration and colonisation. Perhaps modern humans were encountering and interbreeding with Neanderthals along these corridors through central Asia in deep antiquity.

A particular genetic variant associated with fat metabolism, that's curiously prevalent in modern Mexican populations, appears to have come from Neanderthals originally. Perhaps it conferred some sort of advantage in the past, linked to a particular diet, but interacting with a different sort of food intake today it increases the risk of developing diabetes. Other genetic variants that have entered our genomes from these 'lost tribes' are associated with differences in skin and hair colour. Seven out of ten modern Europeans possess a certain gene of Neanderthal origin that's associated with freckles. With other genes inherited from archaic populations, it's less clear what the functional significance is in modern genomes. On the other hand, it's obvious that a lot of archaic DNA has been weeded out – most likely because it was linked to a reduction in fertility.

Interbreeding with lost tribes meant that our ancestors tapped into a rich reservoir of genetic variation – potentially picking up useful adaptations to local environments, including the pathogens that lived in them. This is an important and relatively new insight into the mechanism of evolutionary change: the introduction and spread of a genetic variant might start with a new mutation, or with an old mutation in that population suddenly proving useful – but it can also arrive from another, closely related population, through interbreeding. From apples to humans, we all bear the evidence for our hybrid origins in our genomes.

But it's not just closely related human species we've interbred with that have left their mark on us today. We've found firm allies in other species – including plants as well as animals – and we've met nine of them in this book. By teaming up with these other species, domesticating them – or providing them with an opportunity to 'domesticate themselves' – the course of human history has been profoundly affected, in ways that can be difficult to comprehend. The influence of the Neolithic has rippled down through the centuries and millennia.

The Neolithic Revolution

It's only with hindsight, and with that deep and wide perspective that geography, archaeology, history and genetics provides us with, that we can appreciate this grand narrative. There's such a gulf between mapping out events and processes over thousands of years, on a continental scale, and the personal experiences, the everyday lives, of our ancestors. But on the other hand, it feels as though we're getting closer to a convergence: charred grains, polished stone sickles, traces of milk on fragments of pottery, DNA from ancient wolf bones and echoes of the word for 'apple' in a truly ancient language – each provide us with astonishing glimpses of detail.

Just as we've elaborated the stories of the origin of species, adding more facets and complexity as new evidence comes to light, the story of the Neolithic has become immeasurably more complicated over time. Rather than a linear, predictable march of progress, driven by human intent, developments – new alliances and the new technologies that accompany them – have emerged in a much more haphazard way. The Neolithic – a switch from hunting and gathering nomadism to farming sedentism – was inevitable as human populations grew. But the particular trajectory varied from place to place, and external factors were hugely influential. As the Ice Age loosened its grip on the world, agriculture developed independently in separate areas. Each time it emerged in fits and starts – then the idea, the technology and the newly domesticated species rippled out from their origins, with the power to feed an expanding human population.

The emergence of agriculture in western and eastern Asia near-simultaneously, around 11,000 years ago, must be more than coincidence: global climate change was affecting people – and grasses – separated by thousands of miles. A global increase in atmospheric carbon dioxide levels after 15,000 years ago would have boosted plant production – fields of wild cereals were there for the picking. Then there was the climatic downturn during the Younger Dryas of 12,900 to 11,700 years ago. Hunters would have started to come back empty-handed more often. Easily harvestable fruits and berries would have been thin on the ground. Foragers would have fallen back on their fallback resources – including the difficult-to-collect but energy-rich seeds of grasses: oats, barley, rye and wheat in the west; broomcorn and foxtail millet and rice in the east. Technologies to make harvesting more efficient and to grind the hard seeds into flour, like the Natufian sickles and stone mortars, came long before domestication and farming itself. By the time the climate started improving, this dependence on cereals had developed into proto-farming.

Those early centres of domestication were hugely influential. The broadly Mesopotamian 'cradle of agriculture' provided the founder crops of the western Eurasian Neolithic. From the fertile land between and around the Rivers Euphrates and Tigris came the first domesticated peas, lentils, bitter vetch, chickpeas, flax, barley, emmer wheat and einkorn wheat. From the land around the Yellow and the Yangtze Rivers came millets, rice and soybean. But there were plenty of other locations around the world where domestication took off. At the end of the Younger Dryas, people from the southern half of Africa migrated north to colonise the green, fertile Sahara. They were hunter-gatherers, subsisting on fruits, tubers and cereals as well as the animals they hunted. They'd been using grinding stones since 12,000 years ago, and cultivation of indigenous sorghum and pearl millet may have started there soon after. But Saharan agriculture was wiped out around 5,500 years ago when a southwards shift in the monsoon transformed the once-fertile landscape into a desert. Sugar cane was domesticated in New Guinea some 9,000 years ago, and teosinte was domesticated to become maize, in Mesoamerica, around the same time.

The more we look, it seems the more centres of domestication we find. The Fertile Crescent is fascinating – but it has tended to draw our gaze away from other, equally important wellsprings of the Neolithic. Vavilov identified seven centres of domestication. Jared Diamond posited nine or ten around the world. More recent studies suggest there were as many as twenty-four. Domestication of species has happened many times, in many different places. Quite a few of the environments where domestication took place were – as Vavilov pointed out – mountainous. These are environments where diversity tends to be rich – physical conditions vary with altitude. But for any potential domesticate, the *fit* with human nature, as well as the timing, had to be right. Species that reacted positively to human intervention, at the same time as humans were open to changing their way of life – that was the winning combination which led to the formation of these crucial allegiances. And conscious decision-making rarely played any role.

The term 'artificial selection' perhaps implies an agency, a consciousness, that is not always at work. Although our modern selective-breeding programmes represent carefully planned interventions and incredibly thoughtful selection, it wasn't always so – especially at the dawn of domestication. The wheat that grew up around the threshing floors was not deliberately sown – but it paved the way for the first fields. The separation between natural and artificial selection is, perhaps itself, artificial. Humans are not the only species to affect the evolution of other species. Our very existence depends on interdependency. We might be able to *understand* what it is that we think we've done, peering into genomes, but bees have influenced the evolution of flowers just as surely as we've influenced the evolution of dogs, horses, cattle, rice, wheat and apples. The bees may not know it, and reflect on it as we do, but they have still driven change. What we've called artificial selection, ever since Darwin used the phrase to help build his argument, is no more than human-mediated natural selection.

Domestication may have started, in many cases, as a thoughtless process: species coming into contact, knocking up against each other, growing closer until their evolutionary histories became intertwined. We're so used to thinking of ourselves as the masters, and other species as our willing servants, even our slaves. But the ways in which we entered into these contracts with plants and animals were various and subtle, evolving organically into a state of symbiosis and co-evolution. There was rarely any thoughtful intent behind the initial construction of this partnership. Anthropologists and archaeologists have described three main pathways to domestication of animals – and it was never an 'event', rather a long, drawn-out evolutionary process. One pathway involves animals choosing humans, borrowing resources from us. As they moved in closer, they began to co-evolve with us, becoming tame long before any sort of human-directed selection – like the creation of dog breeds in the last few centuries – could even begin to happen. Dogs and chickens both became our allies in this way. The second route is the prey pathway. Even here, there would have been no initial intention to domesticate animals – only to manage them as a resource. This would have been the route for medium and large herbivores such as sheep, goats and cattle – first hunted as prey, then managed as game, and finally herded as livestock. The final pathway is the most intentional – where humans set out to capture and domesticate animals, right from the start. Usually these animals were seen as useful for something other than just meat – and horses, tamed as our steeds, are a prime example.

Even when conscious intent did start to play a role, as farmers and breeders began to select particular traits – by weeding out the ones they didn't want – the aim still wasn't particularly long-sighted. Darwin himself recognised this. He wrote that, whereas 'eminent breeders try by methodical selection, with a distinct object in view', others would have been focused on just the next generation, with 'no wish or expectation of permanently altering the breed'. Nevertheless, those choices would, over decades and centuries, lead to 'unconscious modification' of a variety or cultivar. Darwin thought that even 'savages' and 'barbarians' (to the modern reader, he's extremely un-PC at times) could modify their animals, by even *less* conscious selection – simply by saving their favoured beasts from being eaten during famines.

The final blow to our mastery of nature comes when we consider the relatively small number of species that we've been able to successfully recruit as our allies. Many, as the nature writer Michael Pollan so succinctly put it, have 'elected to sit it out'. For a species to become a successful ally, it had to possess certain qualities which would – when the occasion presented itself – act as predispositions to becoming human domesticates. Without the curiosity of the wolf, the submissive nature of the mare, the potential for grasses to develop a non-shattering rachis, the plumpness of the central Asian wild apples – we probably wouldn't have dogs, horses, wheat and cultivated apples.

Nevertheless, our domestication of other species has had far-reaching – global – consequences. The concept of interdependency with other species – that lies at the heart of the Neolithic – became an idea, a part of human culture, which would prove so successful that it was destined to spread around the world. Having struck up particular relationships with certain plants and animals, our ancestors could move those species – modifying the local environment to suit them, as they went. It was an extremely successful strategy, even if its origins had come about through pure serendipity.

Today, hunting and gathering is a lifestyle practised by a diminishingly small number of people. Among others, there are still a few, tiny populations of hunter-gatherers in Africa, including the Bushmen in Namibia and the Hadza people of Tanzania. They live in relatively inhospitable places, in semi-deserts – landscapes where farmers can't farm. They've resisted the Neolithic Revolution right up until now, but their way of life is under threat, and will probably disappear this century.

Co-evolution and the course of history

Human history would have played out very differently if the other species we interacted with had been different – missing altogether, impossible to catch or to domesticate, for example. We sometimes approach history and prehistory as though we humans are so much the lords of our own destiny that external forces have little or no role to play. But the story of any species can never be told in isolation. Every species exists in an ecosystem – we are all interlinked and interdependent. And serendipity and contingency are woven into all the interactions that have played out in the course of our intertwined histories.

The alliances that we've formed with other species over millennia have changed the course of human history in ways that the earliest farmers, the first hunters with their dogs, and the first horse-riders couldn't possibly have dreamt of. Cultivated cereals have provided the energy and protein for the expansion of human populations – far beyond the potential that gathered wild foods could support.

The wheats from the centres of domestication in the Middle East provided the fuel for a population boom that spilled over into migrations, with farmers spreading across Europe during the Neolithic. Domesticated sheep, goats and cattle provided that crucial means of storing protein and energy – as 'walking larders'. Forming alliances with plants lately used as fallback foods, and with once-hunted animals, humans began to buffer themselves a little from the immediate effects of climatic disturbances. With a more secure source of energy and protein, and a more settled way of life, families could grow larger. It sounds like such an unmitigated success story, but the slightly counter-intuitive reality of the Neolithic Revolution was that it chained people to lives of hard labour and took its toll on the health of individual women, men and children.

An archaeological site in central Anatolia – spanning just over a millennium, between 9,100 and 8,000 years ago – provides us with an astonishing snapshot of those living through the transition. The early farming community of Çatalhöyük inhabited a dense settlement of mud-brick houses, packed tightly together. At first, just a few families lived there, then the village grew in size dramatically. The farmers grew mainly wheat, but also barley, peas and lentils, and they kept sheep, goats and a few cattle, as well as hunting aurochsen, wild pigs, deer and birds, and gathering wild plants. Their fields were located several miles south of the settlement, and they also had to range widely to hunt and herd their animals. The skeletal remains of more than 600 people have been discovered at Çatalhöyük, and these bones tell tales. There's an extraordinary number of juveniles, including the bones of many newborn babies. On the face of it, it looks like infant and child mortality was particularly high, but in fact this pattern is likely to represent an unusually large number of babies being born in the first place. Teasing apart the numbers by date, the birth rate appears to have risen with the transition from foraging to early farming, and again, with a move to more intensive farming. The number of houses in the village was growing accordingly. Analysis of nitrogen isotopes in the bones of infants indicates that weaning was starting when infants were relatively young, around eighteen months of age. Early weaning in such populations is linked to a shorter space between births – a population boom in the making.

But it wasn't all rosy. Çatalhöyük presents a picture of increased physiological stress and health problems, compared with earlier foraging communities. A diet focused on cereals provides plenty of energy but not necessarily all the essential building blocks of protein, or vitamins, that the body needs. Although other sites have turned up evidence of reduced growth rates, this doesn't seem to have been the case at Çatalhöyük. Nevertheless, there's ample evidence there of low-level physiological stress, including bone infections – and a high rate of tooth decay that is probably linked to a starch-rich diet.

Today, industrialised farming means that the hard graft of agriculture has, to a large extent, become shouldered by machinery rather than humans. But we're all chained to systems of food production where cereals – the fallback foods of our hunter-gatherer ancestors – have become staples, just as they were at Çatalhöyük. With globalised food supplies, we can access other sources of important vitamins (and now we can even insert those into cereals, with gene editing), but our teeth are still suffering from the effects of the Neolithic Revolution. One of the most maligned villains is the sugary derivative of maize: high-fructose corn syrup. It seems to be a food that encapsulates the best and worst of the Neolithic legacy – a fantastic source of energy, certainly, but also an insidious threat to health that we're only just starting to recognise. Maize itself has played a huge part in human history. It fuelled the Inca and Aztec civilisations, and went global after Columbus (and possibly Cabot) reached the New World. Today, we produce more of it by weight than any other grain. It fuels us – but we grow four times as much as we humans eat to feed livestock, and almost as much again to create biofuels.

The impact of domesticated species on our own journey through history is perhaps easiest to grasp if we imagine what would have happened without them. This approach is analogous to the way that geneticists set out to find out about the function of a particular gene – by creating a knockout. We can't test our alternative histories in the same way, but our thought experiments can still give us some idea of how different the world could have been without these various species.

Without our domesticated cereals, where would we be today? The Neolithic would have unfurled in an unfamiliar manner. Pastoralism alone would surely not have supported the population expansions that saw the spread of people, livestock and crops from the Middle East, right across Europe. Would early civilisations – the Sumerian in the Middle East, the Yellow and Yangtze River civilisations in the Far East, and the Mayan in Mesoamerica – have taken off? Perhaps not in the same way, but the horse-riding nomads of the Eurasian Steppe remind us that civilisation can evolve on the move. In a world without cereals, would we all still be nomads, living in yurts rather than houses? Or would starchy tubers like potatoes have filled the void? As we consider the absence of each domesticate, it becomes harder and harder to imagine a world without the species we're so familiar with, and so dependent on.

What about apples? I don't suppose that any civilisation would have crumbled without them, though fruits that can be stored through the winter are few and far between – the lack of apples as a fallback food may have had some impact. We'd still have cider, because we can make that with wild crabapples, and still do. But the wonderful mythology of the apple would be missing from our culture.

Without dogs to help them hunt, perhaps the modern humans of Europe and northern Asia would have been hit even harder by the chilly climax of the last Ice Age, 20,000 years ago. Without wolf-hounds to help us hunt out the last wolves, those predators might just have clung on in Britain and Ireland to the present. Might some of the larger, Ice Age megafauna of Europe have survived to the present day without the lethally effective human–dog alliance pitted against them? Without dogs, perhaps there would even still have been small herds of mammoth roaming in northern Siberia today.

Chickens, as we know, joined the party relatively late – domesticated in the Bronze Age – but then raced into the lead as the most important farmed animals on the planet. Without chickens, we'd never have had the Del Marva Chicken-of-Tomorrow Queen. There would have been no cockfighting. The French football team would've had to come up with a different emblem. And cuisines around the world would miss the chicken's flesh and its egg. There are other domesticated birds, certainly, but none so amenable and successful as the chicken. Of course, that may all change when someone launches the 'Duck-of-Tomorrow' competition.

Imagining how history would have played out without horses is extremely difficult. From the very beginning, domesticated horses would have had a profound economic impact, vastly extending the range over which herders could tend their cattle on the steppe. Would steppe populations have expanded and pushed westwards and eastwards, as they did, without horses? It seems unlikely.

Horses played a vital role in European prehistory. Out of the steppe on the eastern fringes of Europe came the horse-riders, speaking that language that still echoes today. Language wasn't their only contribution; the characteristic mounded-up kurgan and timber grave-burial cultures of Siberia and the Pontic-Caspian Steppe spilled over into Europe, too. Picking up an idea which had evolved on the steppe, Bronze Age people around the eastern Mediterranean – as we've seen – began to bury their kings under huge tumuli, accompanied by luxuries for the afterlife. Those elite burials often contained horse trappings – and sometimes the skeletal remains of horses themselves. The cult of the horse – inextricably bound up with high status in society – continued into the Iron Age and beyond, and continues to echo in the modern world.

Horses were also used for traction. The first wheeled vehicles probably originated in the steppe, and chariots definitely originated there – around 2000 BCE. Then chariots spread east into China and west into Europe. Warfare was transformed when armies started to fight on horseback, in the second millennium BCE, and cavalry continued to be crucial in battles up to and including the First World War. The global history of conflict would have been vastly different without horses. Cattle can be used for draught, but not for cavalry.

Today, horses may have been largely replaced by wheeled vehicles that travel under their own steam (or combustion), but they're still admired and valued for their speed, power and beauty. They continue to be entangled with our ideas of high status. Equestrian sports and blue blood are harnessed together.

The loss of cattle from history may seem as though it would be less impactful, but they've been critical, not just for meat and milk, but for transport and agriculture – pulling carts and ploughs through the centuries. And they've been with us since the early Neolithic – long before horses were domesticated. But, like horses, they became culturally important in a way that transcends their function as beasts of burden and sources of sustenance. Perhaps as such large beasts, in a world where so many megafauna had gone extinct at the end of the Ice Age, they occupied that space in our mythologies. Although domesticated, they symbolised strength, power and danger. The cult of the bull on Crete formed the inspiration for the myth of the minotaur. The mystery religion focused on Mithras, who killed a huge bull, travelled all the way to Britain with the Romans. An image of Mithras is carved into a stone found at one of the forts along Hadrian's Wall. But cattle didn't only find their way into our mythology, they influenced our DNA.

Milk and genes

Although Neolithic cows may have been reared primarily for their meat – remember the riddle of the shrinking cow – the use of milk goes right back to at least the seventh millennium BCE. Milk is a fantastic food: it contains a great range of essential nutrients, including carbohydrate, in the form of lactose, lipids and protein, as well as vitamins and minerals – calcium, magnesium, phosphorus, potassium, selenium and zinc. But it's an unusual food for adult mammals to ingest. Most mammals can't digest milk as adults. It's a characteristic of female mammals to produce milk for their young, and – as mammals – we humans have always been well equipped to drink and digest milk in infancy; we're born expecting to be sustained by our mother's milk. But the ability to digest milk – in particular, the milk sugar lactose – usually disappears by adulthood in mammals: humans included. The gene encoding the necessary enzyme, lactase, gets switched off. And yet, most of us in Europe can happily drink milk into adulthood.

The domestication of cattle (and sheep and goats) has not only affected our history and culture – it's affected our biology as well. By starting to keep animals for milk, we altered our environment. We've certainly altered the DNA of cattle, through that human-mediated natural selection that we tend to call artificial selection – but by drinking milk, we ended up altering the way in which natural selection was acting on *us*. Just as we've been in the business of remaking other species to suit our needs, tastes and desires – they've been remaking us at the same time.

Drinking fresh milk would have presented a real challenge to our ancestors: it would have caused bloating, stomach cramps and diarrhoea in most people brave enough to try it. The problem is caused by that inability to digest lactose, which then stays in the gut and gets fermented by bacteria – leading to all those unpleasant gastrointestinal effects. There is a way of getting around this drawback – and that's to reduce the lactose content of milk. You can do that by fermenting it or turning it into hard cheese, both of which also preserve the milk in potable or edible form for longer.

As Richard Evershed and his team showed – by analysing the lipids on pottery fragments from Poland – Neolithic farmers there were making cheese, probably from cows' milk, as early as the sixth millennium BCE. Mares' milk contains considerably more lactose than cows' milk – but the invention of fermented milk drinks would have transformed mares' milk into something which could be safely drunk by anyone. It's likely that the kumis of the Eurasian Steppe – a mildly alcoholic 'milk-beer', still drunk today – was also a very ancient invention.

But some of us have evolved to be able to drink and digest fresh milk quite comfortably, far beyond those early months when we're dependent on our own mother's milk. We are 'lactose tolerant' – and that comes from possessing an allele, or gene variant, which means we continue producing lactase as adults. The European gene variant linked to lactase persistence is estimated to be around 9,000 years old. In central Europe, early Neolithic populations don't have the variant; it's there at a low frequency by 4,000 years ago, but today up to 98 per cent of people in north-western Europe are lactase persistent (or lactose tolerant). This suggests that their ancestors went through times of drought and strife when the ability to digest fresh milk – not just those stored, fermented milk products and cheeses – may have meant the difference between life and death. The gastrointestinal effects of drinking fresh milk – for people who don't have lactase persistence – were still well known in the first century BCE, when the Roman scholar Varro wrote that mares' milk acted as a good laxative (if that was the effect you were after), followed by donkeys' milk, cows' milk, and finally goats' milk. It seems that lactose tolerance was unusual in Italy even just two thousand years ago. And although it's now very common in western Europe, lactase persistence is only present in around 25–30 per cent of people from Kazakhstan, for instance.

Descendants of dairy farmers in Africa ended up with a similar adaptation, with the African genetic variant originating around 5,000 years ago, then spreading through the population. These dates fit very well with the archaeological evidence for the origin and spread of domesticated cattle. In contrast, most East Asians – without a history of dairy-farming – are unable to drink fresh milk without severe gastrointestinal side effects ensuing.

Lactase persistence is one of the clearest signs of recent adaptation and evolutionary change in human genomes – apart from the many changes relating to disease resistance. While so many people have been seduced into following a 'palaeo' diet, our ancestors' physiology didn't stand still when the Neolithic Revolution transformed ancient ways of life. It's not just the species we've domesticated that have changed – they, in turn, have changed us.

These alliances have started off in different ways. Some may have begun quite inadvertently, like apple pips deposited in middens and growing into new trees. Some may have been instigated by other species – wolves may have initiated the contact that led to some of them becoming domesticated as dogs. Others may have been more deliberate on our part – the catching and taming of horses and cattle surely fall into this category. But regardless of how they started, each alliance developed into a symbiotic ecological relationship – an experiment in co-evolution. Domestication is a two-way process.

But there's another curious link between the animals we've domesticated and ourselves. We too seem to show some of the traits that appeared when animals were domesticated. Like dogs, and Belyaev's silver foxes, we've ended up with smaller jaws and teeth, and flatter faces than our predecessors, as well as diminished male aggression. This suite of associated characteristics has been called the 'domestication syndrome'.

The self-tamed species

Humans are extremely sociable, tolerant creatures. We may sometimes forget this when we look at examples of bad behaviour, on the internet, in politics and even in our daily encounters. Worse, criminality, violence and warfare can make us seem like a hopelessly belligerent species. But history shows that we are, on average, less violent today than we were last century, and the centuries before that. We are learning to live together more peaceably, even if we still have some way to go.

If we compare ourselves with our closest living relatives, chimpanzees and bonobos, we come out extremely well. In other apes, large social groups tend to tear themselves apart, while fear and stress are natural responses to meeting an unfamiliar member of one's own species. Somehow we've managed – most of the time – to endure living in very close proximity to many other humans, to react calmly to encounters with strangers, and to cooperate to an extraordinary extent on shared projects. Indeed, our peculiar success as a species, and the development of our extraordinary cumulative culture, rests on that ability to cooperate and help each other. To achieve that, we've had to become – tame.

Our species emerged at least 300,000 years ago, in Africa. It's likely that the ability for symbolic behaviour, including art and spoken language as forms of communication, was present right from the beginning in *Homo sapiens*, and may even have been present hundreds of thousands of years before, in the common ancestor of us and Neanderthals. After sporadic flashes of symbolic behaviour in the archaeological record – the odd pierced shell, the odd piece of ground-down ochre – it really bursts on to the scene after 50,000 years ago. From then on, we see a great diversity of the types of objects that humans are making; and they begin to make so much art that some of it survives through to the present day – in the form of carved ivory animals and figurines, and painted caves. Anthropological studies into the spread of culture through Tasmania and Oceania hold a clue as to what it was that unlocked all that creativity. If the analogy holds, it's likely that Ice Age culture blossomed as populations reached a large enough size, with enough mobility and connectedness, for ideas to emerge, take hold, spread and evolve.

Still, increases in population density present a particular challenge for any species. More people means more mouths to feed; there's more competition for resources. It's argued that the emergence of 'modern human behaviour' – all that cumulative, cultural complexity – would only have been possible with exceptionally high levels of social tolerance. When we're less afraid, less antagonistic and more open to communication from others – we learn.

Selecting against aggressive tendencies in other animals, from silver foxes to mice, causes widespread changes in behaviour. As you might expect, they become much more friendly. But the changes in behaviour, mediated by hormones, are also accompanied by physical changes – particularly in the shape of the head and face. Tamed silver foxes, as well as displaying patches of white in their coats, have smaller canine teeth and smaller skulls with shorter snouts. The tame adults look like wild juveniles.

Over the last 200,000 years, human skulls have also changed, becoming less robust, with less pronounced brow ridges, thinner bones overall, and less of a difference in the size of canines between males and females. This looks like a similar pattern to that seen in silver foxes and other domesticated animals. The change may be linked to a reduction in testosterone levels – which affects bone growth as well as behaviour. Testosterone produces particular effects at different stages of development. Individuals who've experienced relatively high levels of testosterone in the womb tend to have smaller foreheads, broader faces and more prominent chins. Males with high levels of testosterone at puberty tend to develop a taller face shape and a heavier brow. Men with very 'masculine' faces like this are perceived to be more dominant.

Looking at early fossils of modern humans, they are generally much more heavy-browed than more recent examples, but is it possible to be more specific about when the changes actually took place? A team of evolutionary anthropologists in the US decided to find out, measuring and comparing samples of skulls – some dating to between 200,000 and 90,000 years ago, some from after 80,000 years ago, and a large sample of more recent specimens, dating to within the last 10,000 years. They found that brow ridges were much more pronounced in the skulls dating to more than 90,000 years ago, compared with later samples. The height of the face was also taller in the older sample. The 'feminisation' of face shape continued into the Holocene. It's possible that these changes in face shape were mediated by changes in testosterone levels. If so, then more gracile, feminine skulls – in both sexes – could be a by-product of selection for social tolerance as human populations grew. It's easy to imagine how that selection pressure might have worked. Evolution is, as geneticist Steve Jones has so brilliantly put it, 'an examination with two papers'. It's not enough to simply survive, you must also reproduce – transmitting your genes to the next generation. If you're a social outcast, you may find it difficult to pass, or even to sit, that second paper. If men with reduced levels of aggression had a better chance of sexual success, then that trait would quickly spread through a population. As human society evolved, and our ancestors began to live more densely, as well as relying on extensive social networks to survive, it seems that we may have – quite inadvertently – domesticated *ourselves*.

Domesticated animals share another characteristic with us humans – and we've taken it to an extreme. We tend to develop slowly. We're childish, or puppyish, for longer than our wild counterparts. Infants and juveniles are more trusting, more friendly, more playful, and more receptive to learning than adults. If we think about the various scenarios where animals were either tolerated or caught by humans, and then became not only habituated to human presence but cooperative, then it makes so much more sense if we're talking about juveniles – whether that's puppies, calves or foals. And if, at every generation, those individuals who grew up more slowly, who stayed receptive for longer, were more likely to continue that alliance with humans, we can see how domestication would have – quite inadvertently – exerted a selection pressure to stay 'younger' for longer.

In 'domesticating' ourselves, we've ended up changing how natural selection acts on us too: favouring those who stay youthful – or at least, behave youthfully – for longer. It seems like a simple transformation. Old hypotheses suggested that 'neoteny' was the key here: a sort of arrested development where adult organisms would somehow remain more childlike, both physically and behaviourally. More detailed analyses in biology – and genetics in particular – blows that out of the water. It really never is that simple. Childlike changes are part of it, but they're not the whole explanation. We're really only just beginning to understand the conversations that go on between our genes, our hormones, and our environment – including the other species in it. Nevertheless, there is something that may unite all the changes – neural, physiological and anatomical – which are seen in the 'domestication syndrome' in different animals. That 'something' is a certain population of cells in the embryo which go on to make a great range of tissues in the body – from cells in the adrenal gland to pigment-producing cells in the skin, parts of the facial skeleton and even teeth. The various fates of these embryonic cells – called neural crest cells – seems to map almost too perfectly on to the features of the syndrome. If you had to predict the effects of a defective gene or two involved with neural crest cells, you'd probably say that it would affect particular hormones and behaviour, the shape of the face and the size of teeth, and cause a few interesting changes to pigmentation in the skin. It's just a hypothesis at the moment, but a good one: it makes predictions which are testable. There should be fewer neural crest cells in the embryos of domesticated animals. And if we start to find some mutations linked to domestication that affect neural crest cells in the embryo, then this could explain the basis of the whole syndrome – and why different mammals display similar changes under domestication. Time – and more research – will tell.

The eighteenth-century philosopher Jean-Jacques Rousseau considered civilised humans to be in some way degenerate: a pale, flabby diversion from the original, noble state of the savage. Other, humanist philosophers have seen human 'domestication' as a positive advance, removing us from a more brutal ancestral condition. The discussion around human self-domestication has become mired in political and moral interpretations. Biological ideas are always subject to misuse in this way, but there's no moral dimension to evolution. What happens happens because natural selection favours adaptations that are performing well, at that moment, in that particular environment, and sieves out the rest. What was good for our ancestors may not be so good for us now. They were neither worse nor better than us, from a moral perspective. We got better at living with each other, in close proximity, simply because that *worked*, not because it was morally superior. We wouldn't suggest that the dog is morally superior to the wolf, or the cow to the aurochs, or domesticated wheat to its wild cousins.

The physical changes seen in humans over time, and which seem to reflect a reduction in aggressive tendencies and an increase in tolerance, echo what's seen in domestic animals – but also chime with differences between some wild species. Bonobos are close cousins of chimpanzees – but they're much less aggressive and more playful. Their development, compared with chimpanzees, is also delayed – bonobo infants tend to be less fearful and more dependent on their mothers. There's less difference in skull shape, and in the size of canine teeth, between the sexes in bonobos than there is in chimpanzees. Crucially, these anatomical changes seem to have appeared as incidental by-products of selection for sociability, just as they did in the domesticated silver foxes. It seems that a process akin to 'self-domestication' has actually been quite widespread in mammalian evolution – wherever increasing social tolerance has proven useful to evolutionary success.

While some philosophers have talked about self-domestication in humans representing some sort of escape from the normal rules of evolution, and from natural selection in particular, the presence of a similar suite of characteristics in other – non-domesticated – animals shows this to be quite wrong. Natural selection is still hard at work, even when it's pro-social, non-aggressive, cooperative behaviour that's being selected for. Once again, humans aren't such a special case as we sometimes imagine ourselves to be. Normal rules apply.

And when it comes to the animals we've domesticated, it seems that we may have struck it lucky – we've just harnessed a natural potential by taming those animals, securing them as our allies. That potential may be more developed in some animals than others – depending on how their society and their interactions with members of other species has evolved – perhaps explaining why it's easier to domesticate wolves than wolverines, and horses compared with zebras. And humans – we've always been ripe for self-domestication. Apes are social creatures. We found success in living in denser groups; we became even more sociable. There was no stopping us. We do this puppyish, youthful, playful, trusting thing better than anyone else. And when the Neolithic came along, with the potential to support expanding human populations, our ancestors thrived in that new environment of their own making. As the population boomed, and people started living in closer quarters than ever before, selection for social tolerance would have become even stronger. The people of Çatalhöyük literally lived on top of each other in their small citadel of mud-brick houses. Today, life in huge, dense cities is only possible because we're so socially tolerant, because we've domesticated ourselves. But of course, it's not only *our* environment that we've changed.

The legacy of the Neolithic

Humans exert a profound effect on the physical environment, not just locally, but globally. The conventional view is that anthropogenic climate change – caused by humans – began during the Industrial Revolutions of the eighteenth and nineteenth centuries. Since then, we've been burning fossil fuels in increasing quantities, pushing up the level of carbon dioxide in the atmosphere, and warming the planet. But in fact, our impact on global climate started much earlier – back in the Neolithic. Antarctic ice cores provide a record of ancient levels of carbon dioxide and methane in the atmosphere, and for most of the past 400,000 years the concentrations of these gases has fluctuated in predictable natural cycles. But then the pattern changed – 8,000 years ago for carbon dioxide, and 5,000 years ago for methane. The levels of these gases began to rise when they should have been dropping. The timings correspond with the beginning of the Neolithic in western and eastern Asia, and with the spread and intensification of agriculture. The shift from foraging to farming had a profound impact on the landscape, as forests were cleared to make way for fields – and carbon dioxide was released into the atmosphere. It's possible that this delayed the onset of a glaciation that would otherwise have seen ice sheets descending once more over the Northern Hemisphere. In this period of climatic stability, then, our civilisations grew up and flourished. But now we've undoubtedly gone too far – we're not just tinkering with global climate, we're prodding it hard, and we don't fully understand the long-term consequences of that. If a few thousand humans armed with stone tools could inadvertently warm the climate enough to delay an Ice Age, what damage could more than 7 billion of us do?

Human-induced, anthropogenic climate change represents a clear and present threat, not only to us, but to many other species. But set against that pressing necessity to cut greenhouse-gas emissions is the need to feed a whole world full of people. And our numbers just keep growing. Before the Neolithic, the global human population was just a few million at most. The advent of farming supported a population boom, so that, by 1,000 years ago, it's estimated that there were around 300 million people on the planet. By 1800 that number had risen to a billion.

During the twentieth century, the human population burgeoned from 1.6 billion to 6 billion. Food production needed a huge boost, and got it – in the form of the Green Revolution. Between 1965 and 1985, average crop yields increased by more than 50 per cent. The rate of population growth peaked in the 1960s and is now declining, and the number of humans on the planet seems set to stabilise at around 9 billion, in the middle of this century. But we're still looking at a billion more mouths to feed by 2050. It's enough to spark a mild Malthusian panic.

We seem to need another 'Green Revolution' – but in fact the first one was far from being a sustainable solution: the boost to productivity came at a high cost. Grain for grain, agriculture is now more energy-hungry and more dependent on fossil fuels than it was before that not-so-green revolution. Agriculture is responsible for around a third of global greenhouse-gas emissions – through the clearing of tropical forests, from methane emanating from the rear ends of livestock, as well as that produced by microbes in flooded rice fields, and nitrous oxide wafting up from fertilised soil. There are other problems too: more expensive seeds and a growing emphasis on monoculture and cash crops threaten the livelihoods of poor farmers. Heavy use of agrochemicals has also taken its toll on both human health and wildlife. Changes in land use, together with pesticides, have decimated insect populations. The environmental and health costs of nitrogen contamination from fertilisers are even estimated by some to outweigh the economic gains in agriculture. But perhaps just as importantly, although the Green Revolution boosted food production, it never solved world hunger. This is where it gets insanely complicated, and highly political, because we're already producing enough food for everyone – just not in the right places, or at the right prices. International trade in food generates profits for increasingly large and powerful corporations, but doesn't get food to where it's needed most. There's been a recent expansion of land pressed into agricultural use, but this has been largely to provide meat, oil, sugar, cocoa and coffee for the rich. We're also wasting an obscene amount – a full third of the food we produce. Meanwhile, the poorest people – in both developing and developed countries – still don't have access to the nutritious food they require. Our global food system clearly needs a major overhaul if we're to have a hope of feeding everyone, sustainably.

The key to solving world hunger is unlikely to come simply from driving productivity on large-scale commercial farms – which already produce huge surpluses. Some 90 per cent of the world's farms are smaller than 2 hectares – so supporting smallholders to become more productive is crucial to achieving global food security. Focusing on yield alone is likely to lead to more problems with spiralling energy costs, increased greenhouse-gas emissions, loss of habitats and biodiversity, and contaminated water. Ecologists argue that the best way forward is not through intensification and the use of agrochemicals, but through sustainable 'agro-ecological' methods which are designed to maintain soil and water quality, and to support – rather than to poison – pollinators. We need bees – more than they need us.

GM could be part of the solution. We've seen how a dietary staple could be transformed into something which delivers a much-needed vitamin, in the form of Golden Rice. We now have the tools to make crops that are also better at extracting nutrients, naturally resistant to diseases and drought. We may soon be able to breed flu-resistant chickens and pigs. The prize seems tempting enough – bringing us another step closer to global food security – but of course the technology is still mired in controversy.

Moving parts from one organism to another – including organ transplants in humans – has always caused consternation. In the past, grafting of fruit trees even met with some ethical objections. In the biblical law laid out in the Talmud, dating to the third century BCE, there's a specific prohibition against grafting one type of tree on to another: 'Grafting apple with wild pear, peach with almond or red date with sidr [another date tree], in spite of their similarity, is forbidden.' It was also forbidden to breed together two different kinds of animals. It seems that concern about transgressing the species boundary goes back a long way, and even grafting within a species was condemned by some. The sixteenth-century botanist Jean Ruel called grafting an 'insitione adulteries' – an 'adulterous insertion'. And John 'Johnny Appleseed' Chapman – the man who transported canoe-loads of pips to set up apple-tree nurseries at the frontier in early nineteenth-century North America – railed against the practice. He's quoted as saying: 'They can improve the apple in that way, but that is only a device of man, and it is wicked to cut up trees in that way. The correct method is to select good seeds and plant them in good ground and only God can improve the apple.' There are echoes here with contemporary objections to genetic modification – which, after all, is grafting at a molecular level.

It can be quite easy to fall into the trap of viewing species as monolithic, unchanging. The fact that we don't tend to see one species changing into another over the short time frame of a human life helps to cement that idea. But of course species are *not* immutable. That's the lesson of evolution – which we see in fossils, in the structure of living organisms, and in their DNA. And actually there *are* instances when we do see changes within a human lifetime, or even more quickly. Bacteria reproduce and evolve extremely fast. The appearance and spread of resistance to antibiotics amongst bacteria represents a fast, recent – and hugely troubling – evolutionary change. But it's possible to see 'real-time' evolutionary changes in animals, too, especially where environments have undergone dramatic change – and through selective breeding. Experiments like that of Belyaev with his silver foxes show just how fast those changes can be. And Darwin focused on describing variation and change under domestication in the *Origin* precisely because he knew that this represented evidence of the mutability of species that everyone would be familiar with. Once he'd covered that ground, laying out the evidence for the effects of artificial selection, he could move on to describing how unthinking, natural processes could produce a similar effect: how natural selection could work to create the diversity of life on earth.

A species is in constant flux. Even without novel mutations, the frequency of particular types of gene in a population changes over time: through genetic drift and natural selection – and introduction of DNA from other species. It's the interaction between the members of a species and their environment which produces this dance – some variations do better than others. Mutations, when they do occur, introduce new possibilities into the mix – although they're not the only source of novelty. Sexual reproduction, which involves a shuffling of DNA as gametes are produced, as well as the creation of novel pairs of genes when maternal and paternal chromosomes come together in the fertilised egg, creates variation out of existing genetic material. A changing environment also exerts new pressures. That environment is not only physical but biological – it includes all the other species that an organism interacts with.

We've been influencing the development of our domesticated species through altering their biological and physical environments for centuries. We've moved them around the globe. We've managed the mates they've found to breed with. We've protected them from predators, and ensured that they have a good supply of food. We've affected their DNA profoundly, but everything we've done before (apart from radiation breeding) has been about *indirectly* altering genomes. Gene editing, of course, offers us the potential to *directly* modify genomes.

The recently revealed hybrid nature of so many species, including us – and our domesticated allies – has been a genuine revelation. Even geneticists have been surprised at how permeable the 'species boundary' has turned out to be. It certainly provides us with a novel context for thinking about the ethics of transferring genes from one species to another.

There does seem to be something of a shift occurring within the Green Movement – away from a blanket rejection of genetic modification, and towards the possibility that this technology could represent a useful and environmentally sensitive tool. Tony Juniper, conservation biologist and a former director of Friends of the Earth, has publicly recognised the potential of GM. Speaking on BBC Radio 4's *Today* programme in March 2017, he sounded a cautiously positive note, talking about the potential for using gene-editing techniques to 'accelerate the process of selective breeding', spreading useful alleles within a species. But Juniper was also open to the possibility and potential of some transgenic – between-species – alterations. 'You [could] take genes from the wild relatives of domesticated plants,' he commented, 'and ... apply those into crop varieties more effectively ... helping to solve various problems including climate-change impact, soil damage, [and] water scarcity.' Some people are even starting to talk about 'GM organic'. It would be an extraordinary twist of fate indeed if GM became part of the new, truly Green revolution.

But the ethical considerations around genetic modification go further than the potential biological problems. There's the question of who is carrying out the task, and profiting from it. There are also real concerns about food sovereignty, about forcing a new technology on to people who don't want or need it. On the other hand, pest-resistant Bt Brinjal and vitamin-enriched Golden Rice might offer genuine options to support poor, smallholder farmers. Obstructing those opportunities – especially if that's done without really knowing what the farmers and their communities want – could simply end up preserving the status quo, ensuring that only the rich countries of the Northern hemisphere benefit from new technological advances. A less paternalistic approach, supporting the poorest farmers to make their own, informed choices, seems fairer.

The geneticists at the Roslin Institute – with their research into gene editing in chickens – aren't interested in persuading people to accept this technology, but they'd love them to be better informed, and then to make up their own minds. They're not cultish or evangelical about GM in the least. I think that this is something fundamentally important about the science and technology that's explored and developed in universities – compared with private corporations. There's much less room for vested interests. Most scientists in universities are doing what they do because they believe it's good for humanity, and they tend to be quite self-critical, self-effacing and fairly resistant to exaggerating claims – even when funders encourage them to do so. I'm sure that's a source of intense frustration for the more business-minded and profit-focused managers in our higher-education institutions, but it's absolutely essential. Publicly funded scientists shouldn't be working to maximise profits. They should be free to follow where curiosity leads them, and to explore possibilities which could work for the common good.

None of the geneticists I've met have presented GM as a panacea, but they think it might have some useful applications, and they're keen to work in partnership with farmers in developing countries to explore its usefulness. Mike McGrew at the Roslin Institute was most animated when he was talking about the potential for gene editing in conservation – but he was equally excited by one of his projects in Africa, funded by the Gates Foundation, focusing on improving flocks of chickens in challenging environments. He was also firm about the need to work on this technology in clear sight and through real engagement with communities. He talked about another project he was involved in – trying to make a dairy cow resistant to the parasitic disease, trypanosomiasis, in Africa, moving a gene from another species into the cow. 'You have to tell people what you're planning to do ahead of time and ask if people find that acceptable ... we shouldn't impose our values on other cultures.'

The biggest problem with this new technology, perhaps, is the food sovereignty one. Farming isn't just about food production, it's about power and profit – which are concentrated in the rich North. There's a danger that new GM varieties – however efficient, robust and disease resistant they are – could just cement the inequalities that already exist in the global food system, leaving poor smallholders disenfranchised once again. The first generation of GM crops, like Roundup Ready soy, was largely irrelevant to poor countries – but the second generation could, if not managed well, end up snatching away power and decision-making from poor farmers around the world.

Traditionally – or at least, in the traditions of the last hundred years or so – farmers have been treated as end-users of knowledge – rather than creators of it. This is very different to the way that the Neolithic started – and actually, it's quite distinct from the reality out there, on the ground, in the fields, from the rice terraces of Longsheng to the orchards and pastures of England. Farmers are always innovating, testing possibilities in real-world experiments; they know their land better than anyone else. Research projects which involve farmers right from the beginning reap dividends, and farmers, in turn, are much more likely to adopt innovations that they've helped to develop. Development experts suggest that the whole system needs to be turned on its head: initiatives need to be driven from the grassroots, with national and international support – rather than by the current top-down system of policies, trade agreements and regulation.

It's a tangled, twisted, entwined problem. We need to come up with ways of producing enough food, in the right places, while adapting to climate change and trying not to make it any worse, as well as conserving ecosystems and improving the livelihoods of poor farmers. Whatever the solutions are, they need to be worked out in a connected way. What's really required is an integrated, holistic strategy – but one that looks very carefully at the gains and costs at the local level, as well as the global. If we're to make sensible decisions, for us, for the welfare of our domesticated animals, and for wild species, we need to break away from dichotomies and dogma. It can't be a simple question of industrial, intensive agriculture OR wildlife-friendly small-scale production; the use of agrochemicals OR organic agriculture; a focus on existing varieties OR the creation of new, genetically modified strains. And the solutions will vary from place to place.

So there's global food production, food security, sorted. Apart from – it isn't. Too many people are still starving. We need solutions quickly. And if that wasn't a big enough challenge – what about the rest of life on the planet? What about all the species we haven't domesticated – what about the wilderness? The real legacy of the Neolithic, as far as the planet is concerned, is not how well humans manage to survive and thrive, but how other species around us – the ones we haven't domesticated – have been affected by this revolutionary change.

The wild

I remember flying over Malaysia some ten years ago and feeling heart-wrenching dismay at the extent of the deforestation. Hills and valleys had been scoured clean of their natural vestment of ancient rainforest and bulldozer tracks formed strange, ridged patterns like pink thumbprints on the land. Where green was returning, it was all in orderly rows – of palm-oil seedlings. The monoculture palm-oil plantations covered vast areas, in regular patterns and standard green.

A Malaysian man that I was filming with had links with the palm-oil industry, and I gently expressed my concerns. 'But you deforested your island thousands of years ago,' he said. 'You shouldn't be preaching to us.'

We're currently pushing at the limits of the biosphere to support humans. Some 40 per cent of terrestrial land is farmed – as our population and demand for food grows further, how much more will be pressed into service, under the plough or as pasture for our domesticated herds? Is it at all possible to balance the need to produce our food with the challenge of preserving biodiversity and real wilderness?

Our domesticated livestock – especially large mammals such as cattle, sheep and buffalo – represent a huge burden for the planet. There are over 7 billion of us, and around 20 billion of them. We currently feed a third of the plants we grow to these animals. An increasing volume of the cereals that we grow is being diverted to feed livestock – a topsy-turvy trend that just makes food production even more energy-hungry. We could stop eating meat. At the very least, we could stop eating grain-fed beef and opt for pasture-fed beef instead, or shift from beef to less energy-intensive poultry. We could make our existing food systems much more efficient by making changes like these – without more intensification, without pouring in more energy and agrochemicals. But perhaps we need to consider whether keeping livestock at all can actually be justified any longer. Should we – as a report from the UN's Environment Programme has suggested – be looking at going vegetarian, globally?

Livestock are – rightly – blamed for a litany of ecological problems, but they aren't always detrimental to ecosystems. Sometimes rearing animals can be a way of extracting resources from a landscape which is otherwise difficult to farm – so they're not taking up room that could be used for growing crops. On the other hand, grazing can be disastrous. Writer and environmental activist George Monbiot rather devastatingly called the grazed landscapes of Britain 'sheepwrecked'. But it's not always an unmitigated disaster – carefully managed grazing can help to keep environments like grasslands open, for instance. Having lost so many species of Pleistocene megafauna at the end of the Ice Age, our domestic megafauna can step into that role, by grazing and trampling – helping to sustain communities of plants and animals that thrive in more open environments. Livestock in mixed farming systems can also help to recycle nutrients back into the soil, adding their own natural fertiliser in the form of manure. And very importantly, livestock represent a source of protein and other nutrients that may be difficult to obtain from plants alone, especially in developing countries. Secondary products like leather and wool are important as well, and animals are still used for traction and transport in places where farming is less mechanised. There's also the bond of that 'ancient contract' between people and their domestic animals: a cultural value that is difficult to measure, but that emerges so eloquently in our stories and myths, and which we feel so strongly.

We need to look carefully at how domesticated animals fit into farms of the future. This is a crucial question for society as a whole, and requires careful thought about the values we place on a host of different factors – limiting carbon dioxide emissions versus improving soil quality or conserving an open landscape, for instance. Industrialised systems can be very efficient, but they also rack up some impressive 'food miles' for animal feed, and raise questions about animal welfare. Canadian soil scientist and biologist Henry Janzen suggests that we need to look at each locality, weighing up all the pros and cons, and asking, 'How best do livestock fit here?' Sometimes the answer will be: they don't. But sometimes, it will make a whole lot of sense to be keeping sheep, goats or cattle – these ancient allies of ours – on the land, and we can work to minimise the environmental stresses while embracing the benefits that our cloven-hoofed companions continue to offer us. Keeping livestock *on* the very land they depend on may be more beneficial to both the animals and the ecosystems they interact with.

But just how much room do we allow our farms to take up? This question really centres on whether we choose to make our agricultural land as productive as humanly possible – or aim for more wildlife-friendly farming. Adopting an intensive, 'land-sparing' approach means accepting the loss of wildlife in our fields but, by focusing on agricultural productivity, should allow us to keep more of the wider landscape truly wild. On the face of it, this seems like a sensible option: if we ring-fence our agricultural land and concentrate on making it maximally productive, we can leave plenty of room for wildlife elsewhere. But ecologists have argued that this just doesn't work in the real world. Wild species can't be supported in isolated pockets of habitat. Wildlife – whether it's bees, birds or bears – tends to do better in networks of protected wilderness, semi-natural habitat and managed landscapes. In the UK, natural biodiversity has been profoundly impacted by the intensification of agriculture since the 1960s. Environmentally friendly farms are essential as both refuges and links, with the hedgerows of traditional farmland forming crucial connecting corridors for wildlife. Organic farming, currently accounting for only 1 per cent of agriculture worldwide, supports wild biodiversity, and can be almost as productive as conventional farming, whilst being more profitable. It looks like the most sustainable option, but achieving both food and ecosystem security will involve a range of different approaches, in different places. The 'land-sharing' versus 'land-sparing' debate continues. Treating the choice as a global dichotomy is unhelpful – ecosystems are much more complex than that. Once again, it's a question which needs to start with a local focus, looking carefully at the communities of plants and animals, the opportunities and pressures, in each place.

There's also an important economic imperative to protect wilderness and wildlife – the future of farming depends on it. The process of domestication has, each time, involved sampling the genetic diversity that existed amongst the wild ancestors of domesticates. The DNA of our tamed species often shows clear signs of 'bottlenecks', sometimes associated with original domestication, but also with the narrowing focus of selective breeding over the last couple of centuries, producing the varieties that we grow and rear today. The Green Revolution led to another contraction of diversity, with a concentration on an even narrower portfolio of more productive cultivars. What seems like a neat solution actually presents a significant threat to all our food-production systems. The 'future-proofing' of any ecosystem, any species, lies in the diversity and variation contained within it. We see that in the history of species, of life on earth. If we try to constrain species too much, we seriously limit their potential to adapt to changes in the future – to unusual pathogens as well as to changes in physical environments. The Irish Potato Famine showed just how devastating the effects can be. The wild relatives of our domestic species represent an immense store of genetic and phenotypic variation. Understanding how domestication happened, and tracking down the wild relatives of our tamed species, isn't just interesting from a historical and theoretical perspective. That knowledge and those wild species are important to our modern breeding programmes, to the future of our domesticated plants and animals. If only for very selfish reasons, we need continuing access to that wild library. What's good for the wild species is good for us, too. We're in the same game: evolution, survival. Our own fate is inextricably tied up with the fates of other species.

Wild species are threatened by the presence of our domesticates – at a genetic level. The distinction between domesticated and wild, the human landscape and the natural one, has become increasingly blurred. Genes from domesticated species have already – have always – escaped from our gardens into the wilderness. We're not sure what this introgression of genes from domesticates means for wild species. Natural selection could end up weeding out the 'domestic' genes – it may have already done that; or those genes could be advantageous – and retained. Recent studies have revealed the presence of DNA from many popular domesticated varieties in crabapple genomes. That could significantly affect the future evolution of wild apples – as well as, possibly, diminishing their usefulness for future crop improvement. And even the tightest regulation can't rule out the escape of DNA from genetically modified organisms into wild species.

The genetic connectedness of our domestic species with their wild relatives reminds us what a complex network of relationships we're part of. Our domesticated species have not somehow 'left nature' – they are still part of it. That goes for us, too. We might have a spectacularly profound and far-reaching impact on the rest of the planet, but we're still a biological phenomenon. If anything, the acceptance that we're *part* of nature should encourage us to be more thoughtful about our influence and the way that our existence impacts on other species. We can never stand apart from all those other lives, but we can perhaps push those interactions in a more positive direction. Keeping an eye on the future of farming shouldn't be our only reason for protecting wildlife. We understand the threat that we, as a species, represent to biodiversity. There's a moral obligation to try to balance the basic need to feed and clothe the human population with the need to sustain our fellow species – not just the domesticated ones, but the wild ones too.

We have become a powerful evolutionary force on the planet, shaping landscapes, changing the climate, forming co-evolutionary relationships with other species, and instigating global diasporas of those favoured plants and animals. Through those movements – as much as through any human-mediated natural selection – the genomes of those domesticated species were altered, as they interbred with wild species. Apples still retain the memory of their origin in those wild orchards on the flanks of the Tian Shan mountains, but they're more wild, European crabapple in their genetic make-up. The same goes for domestic pigs, which originated in Anatolia but bred with wild boars as they spread across Europe – until all their mitochondrial DNA signatures were replaced by those of local, wild boars. Horses interbred with their wild counterparts as they galloped away from the steppe. Commercial chickens today have yellow legs – a trait they picked up when ancestral chickens interbred with grey junglefowl in south Asia. This pattern of origin, spread and interbreeding has created such a tangled genetic tapestry in each domesticated species that it's been hard to pick apart the threads. The injection of genes from wild relatives in different places has often led to suggestions of multiple sites of domestication. But as genetics moved on from looking at mitochondrial DNA to exploring whole genomes, and with extraction of ancient DNA from archaeological remains, the real, complicated picture has begun to emerge. It turns out that Vavilov and Darwin were both right. As Vavilov predicted, most domesticated species do appear to have a single, discrete geographic centre of origin. But Darwin was also right about the likelihood of multiple ancestors – not through multiple, separate centres of domestication, but through the hybridisation that happens as the species spreads. Even cattle, purported to have a second centre of domestication, producing zebu cattle, are more likely to have had a single, original centre of domestication in the Near East. Dogs, too, long thought to have originated from two widely separated centres of domestication in Eurasia are – on the basis of the latest analyses – most likely to have emerged from a single origin. Pigs, however, may be a true exception to the rule, with evidence pointing to separate centres of domestication in west and east Eurasia.

We understand domestication much better today than we did just a decade ago. The boundaries we'd drawn between the tamed and the wild back then were too robust and rigid. As we've been unravelling the stories of our allies, we've also illuminated the evolutionary history of our own species. Like them, we're hybrids. Moving around the world, colonising new landscapes, we interbred with our 'wild' relatives, just as horses, cattle, chickens, apples, wheat and rice did too.

Now we're everywhere – and our domesticated species are global phenomena alongside us. It's obvious that the evolutionary success of our domesticates depends a great deal on *us*, but the success of other species which have not been sown, grafted, bred and bridled by us also depends on their ability to survive in a world profoundly influenced by our existence – and that of our domesticated allies. We don't just need to tend the species that have teamed up with us. We need to nurture the untamed wildness – now, more than ever. We can't plough on with the idea that we can separate ourselves from the rest of nature; we need to learn how to live with it. It feels like the challenge of this century is learning how to accept those interrelationships, to thrive with the wildness, not always to fight it.

As I finish writing this book, my apple trees are coming into leaf. I pruned them heavily this year, shaping them to encourage more fruit, but also to please the eye. I stand back from each tree as I'm pruning it – as I would with a painting – testing the balance of the composition, before reaching in again to trim a branch. The blossoms have all gone, and in their place are the small, round, hard beginnings of new apples. They will swell over the coming months until they're ready to be eaten, when the warmth of summer is waning. Underneath the trees, cowslips – carefully mown around – are still nodding their lemon-yellow heads.

Solitary bees are buzzing. Some of the black bullocks in the field beyond the garden crane their heads over the wall to eat the ivy. A great spotted woodpecker shimmies up one of the apple trees, investigating the bark for insect delicacies. There are divisions here, between the wild and the domesticated, the untamed and the tamed. But in the end, it's all one: a tangled bank, beautifully intertwined.

Acknowledgements

Well, what can I say – I am eternally, massively grateful to many kind colleagues and friends who have freely shared their expertise, reading the draft of *Tamed* and offering me their insights, suggestions and amendments. Thank you: Adam Balic, Helen Sang and Mike McGrew at the Roslin Institute, University of Edinburgh (for all your help with chickens and genetics); Ivana Camilleri (for a brief Spanish lesson, including enlightening me on the meaning of sweet *Zorrita*!); Colin Groves, Emeritus Professor at the Australian National University (for endless evolutionary biological wisdom); Laurence Hurst at the University of Bath (for genetic gems and such careful reading; horse Parmesan, anyone?); Nick and Miranda Krestovnikoff (for their wonderful wassailing parties); Greger Larson at the University of Oxford (guru of domestication!); Aoife McLysaght at Trinity College Dublin (for spotting mutations); Mark Pallen at the University of East Anglia and Robin Allaby at the University of Warwick (for sedimentary support); Adam Rutherford (for troubleshooting, early warnings and the occasional jibe, of course); Chris Stringer and Ian Barnes at the Natural History Museum (for much brain-picking at Cheltenham Festival of Science); Bryan Turner at the University of Birmingham (for such attention to detail, down to the molecules of the story); and Catherine Walker (for hot-off-the-press references!). Any mistakes or oversights are of course mine, and mine alone.

Thank you also to my most excellent editor at Hutchinson, Sarah Rigby, and to my fantastically attentive copyeditor Sarah-Jane Forder. I am always grateful for the enormous support and encouragement of my literary editor, Luigi Bonomi, and to the whole, brilliant team at Jo Sarsby Management, who will be helping me take this book on tour.

And thank you, Dave. I know you think this was all your idea, but it really wasn't. OK – maybe just a little bit.

References

Dogs

Arendt, M. *et al.* (2016), 'Diet adaptation in dog reflects spread of prehistoric agriculture', *Heredity*, 117: 301–6.

Botigue, L. R. *et al.* (2016), 'Ancient European dog genomes reveal continuity since the early Neolithic', *BioRxiv*, doi.org/10.1101/068189.

Drake, A. G. *et al.* (2015), '3D morphometric analysis of fossil canid skulls contradicts the suggested domestication of dogs during the late Paleolithic', *Scientific Reports*, 5: 8299.

Druzhkova, A. S. *et al.* (2013), 'Ancient DNA analysis affirms the canid from Altai as a primitive dog', *PLOS ONE*, 8: e57754.

Fan, Z. *et al.* (2016), 'Worldwide patterns of genomic variation and admixture in gray wolves', *Genome Research*, 26: 1–11.

Frantz, L. A. F. *et al.* (2016), 'Genomic and archaeological evidence suggests a dual origin of domestic dogs', *Science*, 352: 1228–31.

Freedman, A. H. *et al.* (2014), 'Genome sequencing highlights the dynamic early history of dogs', *PLOS Genetics*, 10: e1004016.

Freedman, A. H. *et al.* (2016), 'Demographically-based evaluation of genomic regions under selection in domestic dogs', *PLOS Genetics*, 12: e1005851.

Geist, V. (2008), 'When do wolves become dangerous to humans?' www.wisconsinwolffacts.com/forms/geist_2008.pdf

Germonpre, M. *et al.* (2009), 'Fossil dogs and wolves from Palaeolithic sites in Belgium, the Ukraine and Russia: osteometry, ancient DNA and stable isotopes', *Journal of Archaeological Science*, 36: 473–90.

Hindrikson, M. *et al.* (2012), 'Bucking the trend in wolf-dog hybridisation: first evidence from Europe of hybridisation between female dogs and male wolves', *PLOS ONE*, 7: e46465.

Janssens, L. *et al.* (2016), 'The morphology of the mandibular coronoid process does not indicate that Canis lupus chanco is the progenitor to dogs', *Zoomorphology*, 135: 269–77.

Lindblad-Toh, K. *et al.* (2005), 'Genome sequence, comparative analysis and haplotype structure of the domestic dog', *Nature*, 438: 803–19.

Miklosi, A. & Topal, J. (2013), 'What does it take to become "best friends"? Evolutionary changes in canine social competence', *Trends in Cognitive Sciences*, 17: 287–94.

Morey, D. F. & Jeger, R. (2015), 'Palaeolithic dogs: why sustained domestication then?', *Journal of Archaeological Science*, 3: 420–8.

Ovodov, N. D. (2011), 'A 33,000-year-old incipient dog from the Altai Mountains of Siberia: evidence of the earliest domestication disrupted by the last glacial maximum'. *PLOS ONE* 6(7): e22821.

Parker, H. G. *et al.* (2017), 'Genomic analyses reveal the influence of geographic origin, migration and hybridization on modern dog breed development', *Cell Reports*, 19: 697–708.

Reiter, T., Jagoda, E., & Capellini, T. D. (2016), 'Dietary variation and evolution of gene copy number among dog breeds', *PLOS ONE*, 11: e0148899.

Skoglund, P. *et al.* (2015), 'Ancient wolf genome reveals an early divergence of domestic dog ancestors and admixture into high-latitude breeds', *Current Biology*, 25: 1515–19.

Thalmann, O. *et al.* (2013), 'Complete mitochondrial genomes of ancient canids suggest a European origin of domestic dogs', *Science*, 342: 871–4.

Trut, L. *et al.* (2009), 'Animal evolution during domestication: the domesticated fox as a model', *Bioessays*, 31: 349–60.

Wheat

Allaby, R. G. (2015), 'Barley domestication: the end of a central dogma?', *Genome Biology*, 16: 176.

Brown, T. A. *et al.* (2008), 'The complex origins of domesticated crops in the Fertile Crescent', *Trends in Ecology and Evolution*, 24: 103–9.

Comai, L. (2005), 'The advantages and disadvantages of being polyploid', *Nature Reviews Genetics*, 6: 836–46.

Conneller, C. *et al.* (2013), 'Substantial settlement in the European early Mesolithic: new research at Star Carr', *Antiquity*, 86: 1004–20.

Cunniff, J., Charles, M., Jones, G., & Osborne, C. P. (2010), 'Was low atmospheric CO_2 a limiting factor in the origin of agriculture?', *Environmental Archaeology*, 15: 113–23.

Dickson, J. H. *et al.* (2000), 'The omnivorous Tyrolean Iceman: colon contents (meat, cereals, pollen, moss and whipworm) and stable isotope analysis', *Phil. Trans. R. Soc. Lond. B*, 355: 1843–9.

Dietrich, O. *et al.* (2012), 'The role of cult and feasting in the emergence of Neolithic communities. New evidence from Gobekli Tepe, south-eastern Turkey', *Antiquity*, 86: 674–95.

Eitam, D. *et al.* (2015), 'Experimental barley flour production in 12,500-year-old rock-cut mortars in south-western Asia', *PLOS ONE*, 10: e0133306.

Fischer, A. (2003), 'Exchange: artefacts, people and ideas on the move in Mesolithic Europe', in *Mesolithic on the Move*, Larsson, L. *et al.* (eds) Oxbow Books, London.

Fuller, D. Q., Willcox, G., & Allaby, R. G. (2012), 'Early agricultural pathways: moving outside the "core area" hypothesis in south-west Asia', *Journal of Experimental Botany*, 63: 617–33.

Golan, G. *et al.* (2015), 'Genetic evidence for differential selection of grain and embryo weight during wheat evolution under domestication', *Journal of Experimental Botany*, 66: 5703–11.

Killian, B. *et al.* (2007), 'Molecular diversity at 18 loci in 321 wild and domesticate lines reveal no reduction of nucleotide diversity during Triticum monococcum (einkorn) domestication: implications for the origin of agriculture', *Molecular Biology and Evolution*, 24: 2657–68.

Maritime Archaeological Trust (Bouldnor Cliff): http://www.maritimearchaeologytrust.org/bouldnor

Momber, G. *et al.* (2011), 'The Big Dig/Cover Story: Bouldnor Cliff', *British Archaeology*, 121.

Pallen, M. (2015), 'The story behind the paper: sedimentary DNA from a submerged site reveals wheat in the British Isles' *The Microbial Underground*: https://blogs.warwick.ac.uk/microbialunderground/entry/the_story_behind/

Zvelebil, M. (2006), 'Mobility, contact and exchange in the Baltic Sea basin 6000–2000 BC', *Journal of Anthropological Archaeology*, 25: 178–92.

Cattle

Ajmone-Marsan, P. *et al.* (2010), 'On the origin of cattle: how aurochs became cattle and colonised the world', *Evolutionary Anthropology*, 19: 148–57.

Greenfield, H. J. & Arnold, E. R. (2015), '"Go(a)t milk?" New perspectives on the zooarchaeological evidence for the earliest intensification of dairying in south-eastern Europe', *World Archaeology*, 47: 792–818.

Manning, K. *et al.* (2015), 'Size reduction in early European domestic cattle relates to intensification of Neolithic herding strategies', *PLOS ONE*, 10: e0141873.

Meadows, W. C. (ed.), *Through Indian Sign Language: The Fort Sill Ledgers of Hugh Lenox Scott and Iseeo, 1889–1897*, University of Oklahoma Press, Oklahoma 2015.

Prummel,W. & Niekus, M. J. L.Th (2011), 'Late Mesolithic hunting of a small female aurochs in the valley of the River Tjonger (the Netherlands) in the light of Mesolithic aurochs hunting in NW Europe', *Journal of Archaeological Science*, 38: 1456–67.

Roberts, Gordon: http://formby-footprints.co.uk/index.html

Salque, M. *et al.* (2013), 'Earliest evidence for cheese-making in the sixth millennium BC in northern Europe', *Nature*, 493: 522–5.

Singer, M-HS & Gilbert, M. T. P. (2016), 'The draft genome of extinct European aurochs and its implications for de-extinction', *Open Quaternary*, 2: 1–9.

Taberlet, P. *et al.* (2011), 'Conservation genetics of cattle, sheep and goats', *Comptes Rendus Biologies*, 334: 247–54.

Upadhyay, M. R. *et al.* (2017), 'Genetic origin, admixture and populations history of aurochs (Bos primigenius) and primitive European cattle', *Heredity*, 118: 169–76.

Warinner, C. *et al.* (2014), 'Direct evidence of milk consumption from ancient human dental calculus', *Scientific Reports*, 4: 7104.

Maize

Brandolini, A. & Brandolini, A. (2009), 'Maize introduction, evolution and diffusion in Italy', *Maydica*, 54: 233–42.

Desjardins, A. E. & McCarthy, S. A. (2004), 'Milho, makka and yu mai: early journeys of Zea mays to Asia': http://www.nal.usda.gov/research/maize/index.shtml

Doebley, J. (2004), 'The genetics of maize evolution', *Annual Reviews of Genetics*, 38: 37–59.

Gerard, J. & Johnson, T. (1633), *The Herball or Generall Historie of Plantes,* translated by Ollivander, H. & Thomas, H., Velluminous Press, London 2008.

Jones, E. (2006), 'The Matthew of Bristol and the financiers of John Cabot's 1497 voyage to North America', *English Historical Review*, 121: 778–95.

Jones, E. T. (2008), 'Alwyn Ruddock: "John Cabot and the Discovery of America"', *Historical Research*, 81: 224–54.

Matsuoka, Y. *et al.* (2002), 'A single domestication for maize shown by multilocus microsatellite genotyping', *PNAS*, 99: 6080–4.

Mir, C. *et al.* (2013), 'Out of America: tracing the genetic footprints of the global diffusion of maize', *Theoretical and Applied Genetics*, 126: 2671–82.

Piperno, D. R. *et al.* (2009), 'Starch grain and phytolith evidence for early ninth millennium BP maize from the Central Balsas River Valley, Mexico', *PNAS*, 106: 5019–24.

Piperno, D. R. (2015), 'Teosinte before domestication: experimental study of growth and phenotypic variability in late Pleistocene and early Holocene environments', *Quaternary International*, 363: 65–77.

Rebourg, C. *et al.* (2003), 'Maize introduction into Europe: the history reviewed in the light of molecular data', *Theoretical and Applied Genetics*, 106: 895–903.

Tenaillon, M. I. & Charcosset, A. (2011), 'A European perspective on maize history', *Comptes Rendus Biologies*, 334: 221–8.

van Heerwarden, J. *et al.* (2011), 'Genetic signals of origin, spread and introgression in a large sample of maize landraces', *PNAS*, 108: 1088–92.

Potatoes

Ames, M. & Spooner, D. M. (2008), 'DNA from herbarium specimens settles a controversy about the origins of the European potato', *American Journal of Botany*, 95: 252–7.

De Jong, H. (2016), 'Impact of the potato on society', *American Journal of Potato Research*, 93: 415–29.

Dillehay, T. D. *et al.* (2008), 'Monte Verde: seaweed, food, medicine and the peopling of South America', *Science*, 320: 784–6.

Hardy *et al.* (2015), 'The importance of dietary carbohydrate in human evolution', *Quarterly Review of Biology*, 90: 251–68.

Marlowe, F. W. & Berbescue, J. C. (2009), 'Tubers as fallback foods and their impact on Hadza hunter-gatherers', *American Journal of Physical Anthropology*, 40: 751–8.

Sponheimer, M. *et al.* (2013), 'Isotopic evidence of early hominin diets', *PNAS*, 110: 10513–18.

Spooner, D. *et al.* (2012), 'The enigma of Solanum maglia in the origin of the Chilean cultivated potato, Solanum tuberosum Chilotanum group', *Economic Botany*, 66: 12–21.

Spooner, D. M. *et al.* (2014), 'Systematics, diversity, genetics and evolution of wild and cultivated potatoes', *Botanical Review*, 80: 283–383.

Ugent, D. *et al.* (1987), 'Potato remains from a late Pleistocene settlement in south-central Chile', *Economic Botany*, 41: 17–27.

van der Plank, J. E. (1946), 'Origin of the first European potatoes and their reaction to length of day', *Nature*, 3990: 157: 503–5.

Wann, L. S. *et al.* (2015), 'The Tres Ventanas mummies of Peru', *Anatomical Record*, 298: 1026–35.

Chickens

Basheer, A. *et al.* (2015), 'Genetic loci inherited from hens lacking maternal behaviour both inhibit and paradoxically promote this behaviour', *Genet Sel Evol*, 47: 100.

Best, J. & Mulville, J. (2014), 'A bird in the hand: data collation and novel analysis of avian remains from South Uist, Outer Hebrides', *International Journal of Osteoarchaeology*, 24: 384–96.

Bhuiyan, M. S. A. *et al.* (2013), 'Genetic diversity and maternal origin of Bangladeshi chicken', *Molecular Biology and Reproduction*, 40: 4123–8.

Dana, N. *et al.* (2010), 'East Asian contributions to Dutch traditional and western commercial chickens inferred from mtDNA analysis', *Animal Genetics*, 42: 125–33.

Dunn, I. *et al.* (2013), 'Decreased expression of the satiety signal receptor CCKAR is responsible for increased growth and body weight during the domestication of chickens', *Am J Physiol Endocrinol Metab*, 304: E909–E921.

Loog, L. *et al.* (2017), 'Inferring allele frequency trajectories from ancient DNA indicates that selection on a chicken gene coincided with changes in medieval husbandry practices', *Molecular Biology & Evolution*, msx142.

Maltby, M. (1997), 'Domestic fowl on Romano-British sites: inter-site comparisons of abundance', *International Journal of Osteoarchaeology*, 7: 402–14.

Peters, J. *et al.* (2015), 'Questioning new answers regarding Holocene chicken domestication in China', *PNAS*, 112: e2415.

Peters, J. *et al.* (2016), 'Holocene cultural history of red jungle fowl (Gallus gallus) and its domestic descendant in East Asia', *Quaternary Science Review*, 142: 102–19.

Sykes, N. (2012), 'A social perspective on the introduction of exotic animals: the case of the chicken', *World Archaeology*, 44: 158–69.

Thomson, V. A. *et al.* (2014), 'Using ancient DNA to study the origins and dispersal of ancestral Polynesian chickens across the Pacific', *PNAS*, 111: 4826–31

Rice

Bates, J. *et al.* (2016), 'Approaching rice domestication in South Asia: new evidence from Indus settlements in northern India', *Journal of Archaeological Science*, 78: 193–201.

Berleant, R. (2012), 'Beans, peas and rice in the Eastern Caribbean', in *Rice and Beans: A Unique Dish in a Hundred Places*, 81–100. Berg, Oxford.

Choi, J. Y. *et al.* (2017), 'The rice paradox: multiple origins but single domestication in Asian rice', *Molecular Biology & Evolution*, 34: 969–79.

Cohen, D. J. *et al.* (2016), 'The emergence of pottery in China: recent dating of two early pottery cave sites in South China', *Quaternary International*, 441: 36–48.

Crowther, A. *et al.* (2016), 'Ancient crops provide first archaeological signature of the westward Austronesian expansion', *PNAS*, 113: 6635–40.

Dash, S. K. *et al.* (2016), 'High beta-carotene rice in Asia: techniques and implications', *Biofortification of Food Crops*, 26: 359–74.

Fuller, D. Q. *et al.* (2010), 'Consilience of genetics and archaeobotany in the entangled history of rice', *Archaeol Anthropol Sci*, 2: 115–31.

Glover, D. (2010), 'The corporate shaping of GM crops as a technology for the poor', *Journal of Peasant Studies*, 37: 67–90.

Gross, B. L. & Zhao, Z. (2014), 'Archaeological and genetic insights into the origins of domesticated rice', *PNAS*, 111: 6190–7.

Herring, R. & Paarlberg, R. (2016), 'The political economy of biotechnology', *Annu. Rev. Resour. Econ.*, 8: 397–416.

Londo, J. P. *et al.* (2006), 'Phylogeography of Asian wild rice, Oryza rufipogon, reveals multiple independent domestications of cultivated rice, Oryza sativa', *PNAS*, 103: 9578–83.

Mayer, J. E. (2005), 'The Golden Rice controversy: useless science or unfounded criticism?', *Bioscience*, 55: 726–7.

Stone, G. D. (2010), 'The anthropology of genetically modified crops', *Annual Reviews in Anthropology*, 39: 381–400.

Wang, M. *et al.* (2014), 'The genome sequence of African rice (Oryza glaberrima) and evidence for independent domestication', *Nature Genetics*, 9: 982–8.

WHO (2009), *Global prevalence of vitamin A deficiency in populations at risk 1995–2005*: Geneva, World Health Organization.

Wu, X. *et al.* (2012), 'Early pottery at 20,000 years ago in Xianrendong Cave, China', *Science*, 336: 1696–700.

Yang, X. *et al.* (2016), 'New radiocarbon evidence on early rice consumption and farming in south China', *The Holocene*, 1–7.

Zheng, Y. *et al.* (2016), 'Rice domestication revealed by reduced shattering of archaeological rice from the Lower Yangtze Valley', *Nature Scientific Reports*, 6: 28136.

Horses

Bourgeon, L. *et al.* (2017), 'Earliest human presence in North America dated to the last glacial maximum: new radiocarbon dates from Bluefish Caves, Canada', *PLOS ONE*, 12: e0169486.

Cieslak, M. *et al.* (2010), 'Origin and history of mitochondrial DNA lineages in domestic horses', *PLOS ONE*, 5: e15311.

Jonsson, H. *et al.* (2014), 'Speciation with gene flow in equids despite extensive chromosomal plasticity', *PNAS*, 111: 18655–60.

Kooyman, B. *et al.* (2001), 'Identification of horse exploitation by Clovis hunters based on protein analysis', *American Antiquity*, 66: 686–91.

Librado, P. *et al.* (2015), 'Tracking the origins of Yakutian horses and the genetic basis for their fast adaptation to subarctic environments', *PNAS*, E6889–E6897.

Librado, P. *et al.* (2016), 'The evolutionary origin and genetic make-up of domestic horses', *Genetics*, 204: 423–34.

Librado, P. *et al.* (2017), 'Ancient genomic changes associated with domestication of the horse', *Science*, 356: 442–5.

Malavasi, R. & Huber, L. (2016), 'Evidence of heterospecific referential communication from domestic horses (Equus caballus) to humans', *Animal Cognition*, 19: 899–909.

McFadden, B. J. (2005), 'Fossil horses – evidence for evolution', *Science*, 307: 1728–30.

Morey, D. F. & Jeger, R. (2016), 'From wolf to dog: late Pleistocene ecological dynamics, altered trophic strategies, and shifting human perceptions', *Historical Biology*, DOI: 10.1080/08912963.2016.1262854

Orlando, L. *et al.* (2008), 'Ancient DNA clarifies the evolutionary history of American late Pleistocene equids', *Journal of Molecular Evolution*, 66: 533–8.

Orlando, L. *et al.* (2009), 'Revising the recent evolutionary history of equids using ancient DNA', *PNAS*, 106: 21754–9.

Orlando, L. (2015), 'Equids', *Current Biology*, 25: R965–R979.

Outram, A. K. *et al.* (2009), 'The earliest horse harnessing and milking', *Science*, 323: 1332–5.

Owen, R. (1840), 'Fossil Mammalia', in Darwin, D. R. (ed.), *Zoology of the voyage of H.M.S. Beagle, under the command of Captain Fitzroy, during the years 1832 to 1836*, 1(4): 81–111.

Pruvost, M. *et al.* (2011), 'Genotypes of predomestic horses match phenotypes painted in Palaeolithic works of cave art', *PNAS*, 108: 18626–30.

Smith, A. V. *et al.* (2016), 'Functionally relevant responses to human facial expressions of emotion in the domestic horse (Equus caballus)', *Biology Letters*, 12: 20150907.

Sommer, R. S. *et al.* (2011), 'Holocene survival of the wild horse in Europe: a matter of open landscape?', *Journal of Quaternary Science*, 26: 805–12.

Vila, C. *et al.* (2001), 'Widespread origins of domestic horse lineages', *Science*, 291: 474–7.

Vilstrup, J. T. *et al.* (2013), 'Mitochondrial phylogenomics of modern and ancient equids', *PLOS ONE*, 8: e55950.

Waters, M. R. *et al.* (2015), 'Late Pleistocene horse and camel hunting at the southern margin of the ice-free corridor: reassessing the age of Wally's Beach, Canada', *PNAS*, 112: 4263–7.

Wendle, J. (2016), 'Animals rule Chernobyl 30 years after nuclear disaster', *National Geographic*, 18 April 2016.

Xia, C. *et al.* (2014), 'Reintroduction of Przewalski's horse (Equus ferus przewalskii) in Xinjiang, China: the status and experience', *Biological Conservation*, 177: 142–7.

Yang, Y. *et al.* (2017), 'The origin of Chinese domestic horses revealed with novel mtDNA variants', *Animal Science Journal*, 88: 19–26.

Apples

Adams, S. (1994), 'Roots: returning to the apple's birthplace', *Agricultural Research*, November 1994: 18–21.

Coart, E *et al.* (2006), 'Chloroplast diversity in the genus *Malus*: new insights into the relationship between the European wild apple (*Malus sylvestris* (L.) Mill.) and the domesticated apple (*Malus domestica* Borkh.), *Molecular Ecology*, 15: 2171–82.

Cornille, A. *et al.* (2012), 'New insight into the history of domesticated apple: secondary contribution of the European wild apple to the genome of cultivated varieties', *PLOS Genetics*, 8: e1002703.

Cornille, A. *et al*. (2014), 'The domestication and evolutionary ecology of apples', *Trends in Genetics*, 30: 57–65.

Harris, S. A., Robinson, J. P., & Juniper, B. E. (2002), 'Genetic clues to the origin of the apple', *Trends in Genetics*, 18: 426–30.

Homer, *The Odyssey*, translated by Robert Fagles, Penguin: London, 1996.

Juniper, B. E. & Mabberley, D. J., *The Story of the Apple*, Timber Press: Portland, Oregon, 2006.

Khan, M. A. *et al*. (2014), 'Fruit quality traits have played critical roles in domestication of the apple'.*The Plant Genome*, 7: 1–18.

Motuzaite Matuzeviciute, G. *et al*. (2017), 'Ecology and subsistence at the Mesolithic and Bronze Age site of Aigyrzhal-2, Naryn Valley, Kyrgyzstan', *Quaternary International*, 437: 35–49.

Mudge, K. *et al*. (2009), 'A history of grafting', *Horticultural Reviews*, 35: 437–93.

Spengler, R. *et al*. (2014), 'Early agriculture and crop transmission among Bronze Age mobile pastoralists of central Asia', *Proc. R. Soc. B*, 281: 20133382.

Volk, G. M. *et al*. (2015), 'The vulnerability of US apple (Malus) genetic resources', *Genetic Resources in Crop Evolution*, 62: 765–94.

Humans

Abi-Rached, L. *et al*. (2011), 'The shaping of modern human immune systems by multiregional admixture with archaic humans', *Science*, 334: 89–94.

Benton, T. (2016), 'The many faces of food security', *International Affairs*, 6: 1505–15.

Bogh, M. K. B. *et al*. (2010), 'Vitamin D production after UVB exposure depends on baseline vitamin D and total cholesterol but not on skin pigmentation', *Journal of Investigative Dermatology*, 130: 546–53.

Brune, M. (2007), 'On human self-domestication, psychiatry and eugenics', *Philosophy, Ethics and Humanities in Medicine*, 2: 21.

Cieri, R. L. *et al*. (2014), 'Craniofacial feminization, social tolerance and the origins of behavioural modernity', *Current Anthropology*, 55: 419–43.

Elias, P. M., Williams, M. L., & Bikle, D. D. (2016), 'The vitamin D hypothesis: dead or alive?', *American Journal of Physical Anthropology*, 161: 756–7.

Fan, S. *et al*. (2016), 'Going global by adapting local: a review of recent human adaptation', *Science*, 354: 54–8.

Gibbons, A. (2014), 'How we tamed ourselves – and became modern', *Science*, 346: 405–6.

Hare, B., Wobber, V., & Wrangham, R. (2012), 'The self-domestication hypothesis: evolution of bonobo psychology is due to selection against aggression', *Animal Behaviour*, 83: 573–85.

Hertwich, E. G. *et al.* (2010), *Assessing the environmental impacts of consumption and production*, UNEP International Panel for Sustainable Resource Management.

Hublin, J-J, *et al.* (2017) New fossils from Jebel Irhoud, Morocco and the pan-African origin of *Homo sapiens*. *Nature*, 546: 289–92.

Janzen, H. H. (2011), 'What place for livestock on a re-greening earth?', *Animal Feed Science and Technology*, 166–7; 783–96.

Jones, S., *Almost Like a Whale*, Black Swan: London, 2000.

Larsen, C. S. *et al.* (2015), 'Bioarchaeology of Neolithic Çatalhöyük: lives and lifestyles of an early farming society in transition', *Journal of World Prehistory*, 28: 27–68.

Larson, G. & Burger, J. (2013), 'A population genetics view of animal domestication', *Trends in Genetics*, 29: 197–205.

Larson, G. & Fuller, D. Q. (2014), 'The evolution of animal domestication', *Annu. Rev. Ecol. Evol. Syst.*, 45: 115–36.

Macmillan, T. & Benton, T. G. (2014), 'Engage farmers in research', *Nature*, 509: 25–7.

Nair-Shalliker, V. *et al.* (2013), 'Personal sun exposure and serum 25-hydroxy vitamin D concentrations', *Photochemistry and Photobiology*, 89: 208–14.

Nielsen, R. *et al.* (2017), 'Tracing the peopling of the world through genomics', *Nature*, 541: 302–10.

Racimo, F. *et al.* (2015), 'Evidence for archaic adaptive introgression in humans', *Nature Reviews: Genetics*, 16: 359–71.

Reganold, J. P. & Wachter, J. M. (2016), 'Organic agriculture in the twenty-first century', *Nature Plants*, 2: 1–8.

Rowley-Conwy, P. (2011), 'Westward Ho! The spread of agriculture from central Europe to the Atlantic', *Current Anthropology*, 52: S431–S451.

Ruddiman, W. F. (2005), 'How did humans first alter global climate?', *Scientific American*, 292: 46–53.

Schlebusch, C. M., *et al.* (2017) Ancient genomes from southern Africa pushes modern human divergence beyond 260,000 years ago. *BioRxiv* DOI: 10.1101/145409

Stringer, C. & Galway-Witham, J. (2017) On the origin of our species. *Nature*, 546: 212–14.

Tscharntke, T. *et al.* (2012), 'Global food security, biodiversity conservation and the future of agricultural intensification', *Biological Conservation*, 151: 53–9.

Wallace, G. R., Roberts, A. M., Smith, R. L., & Moots, R. J. (2015), 'A Darwinian view of Behcet's disease', *Investigative Ophthalmology and Visual Science*, 56: 1717.

Whitfield, S. *et al.* (2015), 'Sustainability spaces for complex agri-food systems', *Food Security*, 7: 1291–7.

Index